T0329355

Day-by-Day Math Thinking Routines in Fifth Grade

Day-by-Day Math Thinking Routines in Fifth Grade helps you provide students with a review of the foundational ideas in math, every day of the week! Based on the bestselling *Daily Math Thinking Routines in Action*, the book follows the simple premise that frequent, rigorous, engaging practice leads to mastery and retention of concepts, ideas, and skills. These worksheet-free, academically rigorous routines and prompts follow grade level priority standards and include whole group, individual, and partner work. The book can be used with any math program, or for small groups, workstations, or homework.

Inside you will find:

- 40 weeks of practice
- 1 activity a day
- 200 activities total
- Answer Key

For each week, the Anchor Routines cover these key areas: Monday: General Thinking Routines; Tuesday: Vocabulary; Wednesday: Place Value; Thursday: Fluency; and Friday: Problem Solving. Get your students' math muscles moving with the easy-to-follow routines in this book!

Dr. Nicki Newton has been an educator for over 30 years, working both nationally and internationally with students of all ages. She has worked on developing Math Workshop and Guided Math Institutes around the country; visit her website at www.drnickinewton.com. She is also an avid blogger (www.guidedmath.wordpress.com), tweeter (@drnickimath) and Pinterest pinner (www.pinterest.com/drnicki7).

Dawn Schafer has more than 20 years of mathematics education experience as a middle school teacher in both California and New York, an instructional and content coach, a professional development facilitator, and a support provider for the National Board for Professional Teaching Standards candidates. Most recently, she has been working with teachers to create curriculum focused on inquiry, collaboration, and discovery. She believes that nurturing a growth mindset and prioritizing process over product can help students understand and unlock their true potential as mathematicians.

Also Available from Dr. Nicki Newton
(www.routledge.com/K-12)

Day-by-Day Math Thinking Routines in Kindergarten:
40 Weeks of Quick Prompts and Activities

Day-by-Day Math Thinking Routines in First Grade:
40 Weeks of Quick Prompts and Activities

Day-by-Day Math Thinking Routines in Second Grade:
40 Weeks of Quick Prompts and Activities

Day-by-Day Math Thinking Routines in Third Grade:
40 Weeks of Quick Prompts and Activities

Day-by-Day Math Thinking Routines in Fourth Grade:
40 Weeks of Quick Prompts and Activities

Daily Math Thinking Routines in Action:
Distributed Practices Across the Year

Leveling Math Workstations in Grades K–2:
Strategies for Differentiated Practice

Fluency Doesn't Just Happen with Addition and Subtraction:
Strategies and Models for Teaching the Basic Facts
With Ann Elise Record and Alison J. Mello

Mathematizing Your School:
Creating a Culture for Math Success
Co-authored by Janet Nuzzie

Math Problem Solving in Action:
Getting Students to Love Word Problems, Grades K–2

Math Problem Solving in Action:
Getting Students to Love Word Problems, Grades 3–5

Guided Math in Action:
Building Each Student's Mathematical Proficiency with Small-Group Instruction

Math Workshop in Action:
Strategies for Grades K–5

Math Running Records in Action:
A Framework for Assessing Basic Fact Fluency in Grades K–5

Math Workstations in Action:
Powerful Possibilities for Engaged Learning in Grades 3–5

Day-by-Day Math Thinking Routines in Fifth Grade

40 Weeks of Quick Prompts and Activities

Dr. Nicki Newton and Dawn Schafer

NEW YORK AND LONDON

First published 2021
by Routledge
52 Vanderbilt Avenue, New York, NY 10017

and by Routledge
2 Park Square, Milton Park, Abingdon, Oxon, OX14 4RN

Routledge is an imprint of the Taylor & Francis Group, an informa business

Library of Congress Cataloging-in-Publication Data
Names: Newton, Nicki, author.
Title: Day-by-day math thinking routines in fifth grade: 40 weeks of quick prompts and activities / Dr. Nicki Newton.
Description: 1. | New York : Routledge, 2020. |
Identifiers: LCCN 2019051703 (print) | LCCN 2019051704 (ebook) | ISBN 9780367901776 (hardback) | ISBN 9780367901769 (paperback) | ISBN 9781003022930 (ebook)
Subjects: LCSH: Mathematics—Study and teaching (Elementary)—Activity programs.
Classification: LCC QA135.6 .N48455 2020 (print) | LCC QA135.6 (ebook) | DDC 372.7/049--dc23
LC record available at https://lccn.loc.gov/2019051703
LC ebook record available at https://lccn.loc.gov/2019051704

ISBN: 978-0-367-90177-6 (hbk)
ISBN: 978-0-367-90176-9 (pbk)
ISBN: 978-1-003-02293-0 (ebk)

Contents

Acknowledgments

Thank you to Dawn Schafer for co-writing this book with me.

Meet the Authors

Dr. Nicki Newton has been an educator for over 30 years, working both nationally and internationally, with students of all ages. Having spent the first part of her career as a literacy and social studies specialist, she built on those frameworks to inform her math work. She believes that math is intricately intertwined with reading, writing, listening and speaking. She has worked on developing Math Workshop and Guided Math Institutes around the country. Most recently, she has been helping districts and schools nationwide to integrate their State Standards for Mathematics and think deeply about how to teach these within a Math Workshop Model. Dr. Nicki works with teachers, coaches and administrators to make math come alive by considering the powerful impact of building a community of mathematicians who make meaning of real math together. When students do real math, they learn it. They own it, they understand it, and they can do it. Every one of them. Dr. Nicki is also an avid blogger (www.guidedmath.wordpress.com) and Pinterest pinner (https://www.pinterest.com/drnicki7/).

Dawn Schafer has more than 20 years of mathematics education experience as a middle school teacher in both California and New York, an instructional and content coach, a professional development facilitator, and a support provider for the National Board for Professional Teaching Standards candidates. Most recently, she has been working with teachers to create curriculum focused on inquiry, collaboration, and discovery. She believes that nurturing a growth mindset and prioritizing process over product can help students understand and unlock their true potential as mathematicians.

Introduction

Welcome to this exciting new series of daily math thinking routines. I have been doing thinking routines for years. People ask me all the time if I have these written down somewhere. So, I wrote a book. Now, that has turned into a grade level series so that people can do them with prompts that address their grade level standards. This is the anti-worksheet workbook!

The goal is to get students reflecting on their thinking and communicating their mathematical thinking with partners and the whole class about the math they are learning. Marzano (2007)[1] notes that

> initial understanding, albeit a good one, does not suffice for learning that is aimed at long-term retention and use of knowledge. Rather, students must have opportunities to practice new skills and deepen their understanding of new information. Without this type of extended processing, knowledge that students initially understand might fade and be lost over time.

The daily math thinking routines in this book focus on taking Depth of Knowledge activity level 1 activities, to DOK level 2 and 3 activities. Many of the questions are open. For example, we turn the traditional rounding question on its head. Instead of telling students "Round .54 to the nearest tenth." Inspired by Marion Smalls (2009)[2] we ask, "What are 3 decimals that when rounded to the nearest tenth are .5?"

In this series, we mainly work on priority standards, although we do address some of the supporting and additional standards. This book is not intended to cover every standard. Rather it is meant to provide ongoing daily review of the foundational ideas in math. There is a focus for each day of the week.

- Monday: General Thinking Routines
- Tuesday: Vocabulary
- Wednesday: Place Value
- Thursday: Fluency (American and British Number Talks, Number Strings)
- Friday: Problem Solving

On Monday the focus is on general daily thinking routines (What Doesn't Belong?, True or False?, Convince Me!), that review various priority standards from the different domains (Geometry, Algebraic Thinking, Counting, Measurement, Number Sense). Every Tuesday there is an emphasis on Vocabulary because math is a language and if you don't know the words then you can't speak it. There is a continuous review of foundational words through different games (Tic Tac Toe, Match, Bingo), because students need at least 6 encounters with a word to own it. On Wednesday there is often an emphasis on place value. Thursday is always some sort of fluency routine (American or British Number Talks and Number Strings). Finally, Fridays are Problem Solving routines.

The book starts with a review of fourth grade priority standards and then as the weeks progress the current grade level standards are integrated throughout. There is a heavy emphasis on multidigit operations, fractions and decimals. There is also an emphasis on geometry concepts and some data and measurement. There are various opportunities to work with word problems throughout the year.

1 Marzano, R. J. (2007). *The art and science of teaching: A comprehensive framework for effective instruction*. ASCD: Virginia.
2 Small, M. (2009). *Good questions: Great ways to differentiate mathematics instruction*. Teachers College Press: New York.

Throughout the book there is an emphasis on the mathematical practices/processes (SMP, 2010[3]; NCTM, 2000[4]). Students are expected to problem solve in different ways. They are expected to reason by contextualizing and decontextualizing numbers. They are expected to communicate their thinking to partners and the whole group using the precise mathematical vocabulary. Part of this is reading the work of others, listening to others' explanations, writing about their work and then speaking about their work and the work of others in respectful ways. Students are expected to model their thinking with tools and templates. Students are continuously asked to think about the pattern and structure of numbers as they work through the activities.

These activities focus on building mathematical proficiency as defined by the NAP 2001[5]. These activities focus on conceptual understanding, procedural fluency, adaptive reasoning, strategic competence and a student's mathematical disposition. This book can be used with any math program. These are jump starters to the day. They are getting the math muscle moving at the beginning of the day.

Math routines are a form of "guided practice." Marzano (2007) notes that although the:

> guided practice is the place where students—working alone, with other students, or with the teacher—engage in the cognitive processing activities of organizing, reviewing, rehearsing, summarizing, comparing, and contrasting. However, it is important that all students engage in these activities. (Rosenshine, cited p.7 in Marzano, 2007)

These are engaging, standards-based, academically rigorous activities that provide meaningful routines that develop mathematical proficiency. The work can also be used for practice within small groups, workstations and also sent home home as questions for homework.

We have focused on coherence from grade to grade, rigor of thinking, and focus on understanding and being able to explain the math the students are doing. We have intended to take deeper dives into the math, not rushing to the topics of the next grade but going deeper into the topics of the grade at hand. Here is our criteria for selecting the routines:

- Engaging
- Easy to learn
- Repeatable
- Open-ended
- Easy to differentiate (adapt and extend for different levels).

3 The Standards of Mathematical Practice. "Common Core State Standards for Mathematical Practice." Washington, D.C.: National Governors Association Center for Best Practices, Council of Chief State School Officers, 2010. Retrieved on December 1, 2019 from: www.corestandards.org/Math/Practice.

4 National Council of Teachers of Mathematics. (2000). *Principles and standards for school mathematics*. Reston, VA: National Council of Teachers of Mathematics.

5 Kilpatrick, J., Swafford, J., and Findell, B. (eds.) (2001). *Adding it up: Helping children learn mathematics*. Washington, DC: National Academy Press.

Figure 1.1 Talking about the Routine!

Monday: What Doesn't Belong?

Look at the boxes. Pick the one that doesn't belong.

$(2 \times 3) + (2 \times 3)$	$48 \div 4$
$108 \div 9$	$20 - 9$

pints	liters
kilogram	gallons

Tuesday: Vocabulary Tic Tac Toe

Play rock, paper, scissors to see who goes first. Then take turns, picking a square, saying what it means, drawing or writing something about the word on the side and then marking the word with an x or an o. Whoever gets 3 in a row first wins.

Game 1: Draw it or write the definition

quotient	dividend	difference
product	centimeter	tape diagram
multiple	area	open number line

Game 2: Draw it or write the definition

Draw a shape with 1 right angle.	Draw a parallelogram.	Draw a quadrilateral.
Draw a rectangular prism.	Draw a hexagon that does not look like this ⬢	Draw a rhombus.
Draw a shape with 1 pair of parallel sides.	Draw some examples of polygons.	Draw a kite.

Wednesday: 3 Truths and a Fib

Three of the problems are correct and 1 is false in each set. Find the false one.

A.	B.
1. $\frac{.25}{.05} = 5$	1. $\frac{1}{3} = \frac{2}{6}$
2. $\frac{.75}{.25} = 3$	2. $\frac{4}{5} = \frac{8}{20}$
3. $\frac{1}{.25} = 4$	3. $\frac{2}{5} = \frac{4}{10}$
4. $\frac{1}{.20} = 4$	4. $\frac{1}{4} = \frac{2}{8}$

Thursday: Number Talk

Multiply 2 numbers that have a product near 150.

Friday: Perimeter Problem

Mary bought a rug with an area of 24 square feet. What could the perimeter have been?

Figure 1.2 The Math Routine Cycle of Engagement

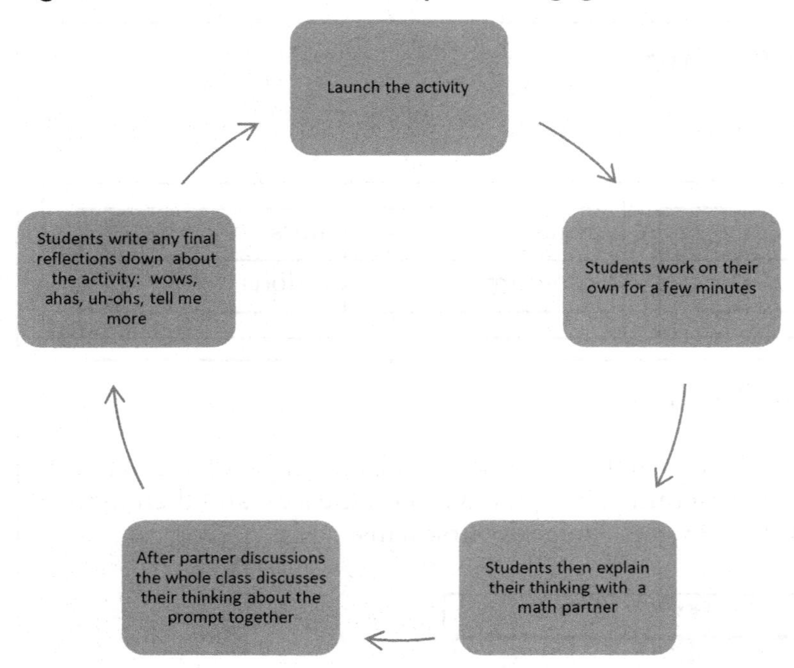

Step 1: Students are given the launch prompt. The teacher explains the prompt and makes sure that everyone understands what they are working on.

Step 2: They are given a few minutes to work on that prompt by themselves.

Step 3: The next step is for students to work with a math partner. As they work with this partner, students are expected to listen to what their partner did as well as explain their own work.

Step 4: Students come back together as a whole group and discuss the math. They are encouraged to talk about how they solved it and the similarities and differences between their thinking and their partner's thinking.

Step 5: Students reflect on the prompt of the day, thinking about what wowed them, what made them say "Ah-ha", what made them say "Uh-oh", what made them say, "I need to know more about this."

Thinking Activities

These are carefully planned practice activities to get students to think. They are **not meant to be used as a workbook**. This is a thinking activity book. The emphasis is on students doing their own work, explaining what they did with a partner and then sharing out to the entire class.

Overview of the Routines

Monday Routines – General Review (Algebraic Thinking, Measurement, Data, Geometry)

- Alike and Different
- Always, Sometimes, Never
- Convince Me!
- Coordinate Grid
- Find and Fix the Error
- It Is/ It Isn't
- Input/Output Table
- Legs and Feet
- Magic Square
- Missing Number
- Number Line It!
- Open Array Puzzle
- Patterns/Skip Counting
- Reasoning Matrices
- 3 Truths and a Fib
- True or False?
- 2 Arguments
- What Doesn't Belong?
- What's the Problem?
- Why Is It Not?

Tuesday Routines – Vocabulary

- Alike and Different
- Convince Me!
- Frayer Model
- It Is/It Isn't
- 1-Minute Essay
- Talk and Draw
- Vocabulary Bingo
- Vocabulary Brainstorm
- Vocabulary Fill-in
- Vocabulary Talk
- Vocabulary Tic Tac Toe
- Vocabulary Match
- What Doesn't Belong?

Wednesday Routines – Place Value/Fractions

- Convince Me!
- Decimal Comparison
- Decimal of the Day
- Find and Fix the Error
- Fraction of the Day

- Fraction Bingo
- Greater Than, Less Than, in Between
- Guess My Number
- How Many More to
- Input/Output Table
- Money Mix Up
- Number Bond It!
- Number Line It!
- Number of the Day
- Place Value Puzzle
- Rounding
- Patterns/Skip Counting
- Start With… Get to … By
- 10 Times as Much
- 3 Truths and a Fib
- True or False?
- Venn Diagram
- Vocabulary Fill-in
- What Doesn't Belong?
- Why Is It Not?

Thursday Routines – Number Talks

- British Number Talk
- Find and Fix the Error
- Number Talk
- Number Talk Puzzle
- Number Strings

Friday Routines – Problem Solving

- Elapsed Time
- Equation Match
- Fill in the Problem!
- Get Close to
- Graph Talk
- Make Your Own Problem!
- Model That!
- Picture That!
- Perimeter Problem
- What's the Problem?
- What's the Question? (3 Read Protocol)
- What's the Story? Here's the Model
- Word Problem Fill-in

Figure 1.3 Overview of the Routines

Routine	Purpose	Description
Alike and Different	This routine focuses on students discussing how numbers and figures are alike and different using the appropriate vocabulary.	Students are given 2 or more of something to discuss. They talk about how they are alike and different.
Always, Sometimes, Never	This routine focuses on students reasoning about whether a statement is always, sometimes or never true.	In this routine, students are given a statement and they have to argue and prove their thinking about if the statement is always, sometimes or never true.
British Number Talk	This routine focuses on students thinking about their thinking.	Students have to choose their own problems and discuss how they are going to solve them. They must name the way they did it, either in their head, with a model or with an algorithm.
Coordinate Grid	This routine focuses on students reasoning about the coordinate grid.	In this routine, students have to plot numbers and discuss how they know they are correct.
Convince Me!	This routine focuses on students reasoning about different topics. They have to convince their peers about specific statements.	Students are given different things to think about like statements or equations and they have to convince their peers that they are correct.
Decimal Comparison	This routine focuses on comparing decimals with hundredth grids.	Students have to choose 2 decimals, model them on the grid and tell a comparison story.
Decimal of the Day	This routine focuses on reviewing different parts of decimals.	Students are asked to write the decimal in word form, expanded form, on the number line, to compare them etc.
Elapsed Time	This routine is an open question, where students work with elapsed time.	Students have to write an elapsed time problem.
Equation Match	This routine focuses on students thinking about which operation they would use to solve a problem. It requires that they reason about the actions that are happening in the problem and then what they are required to do to solve the problem.	Students are trying to pair the word problem and the equation.
Fill in the Problem!	Students have to reason about numbers.	Students are asked to fill in numbers to make a word problem.

Routine	Purpose	Description
Find and Fix the Error	This routine requires that students analyze the work of others and discuss what went well or what went wrong. The purpose of the routine is not only to get students to identify common errors but also to get them to justify their own thinking about the problem.	Students think about a problem either by themselves, with a partner or with the whole group that is either done correctly or incorrectly. They have to figure out why it is done incorrectly or correctly and discuss.
Fraction Bingo	This routine requires students to reason about fractions.	Students play bingo. The calling cards emphasize different fraction concepts, like comparison, size, place on the number line etc.
Fraction of the Day	This routine focuses on students thinking about and modeling fractions.	Students are given a fraction and they have to write it in word form, draw a visual model, plot it on a number line and discuss it in relationship to other fractions.
Frayer Model	This routine is meant to get students talking about concepts. They are supposed to talk about the definition, what happens in real life, sketch an example and also give nonexamples.	Students are given a template with labels. They work through the template writing and drawing about the specified topic.
Get Close to	This routine allows students to think about different ways to get close to a particular number.	Students are given a number and they have to come up with an operation that gets them close to the number.
Graph Talk	This routine gives students the opportunity to tell a story about a graph.	Students are given a graph and they have to tell a story and write questions about the information in the graph.
Guess My Number/ Fraction	This routine gives students a variety of clues about a number or fraction and asks the students to guess which number or fraction it might be given all the clues. Students have to use their understanding of place value and math vocabulary to figure out which number or fraction is being discussed.	Students are given various clues about a number or fraction and they must use the clues to guess which number or fraction it is.
Greater Than, Less Than, in Between	In this routine, students specifically talk about numbers in terms of greater than, less than and in between each other.	Students are thinking about the number relationships and filling in boxes based on those relationships.

Routine	Purpose	Description
How Many More to	In this routine, students are asked to tell how many more to a specific number. Again, this is another place value routine, asking students to reason about numbers on the number line.	Students are given a specified number and they have to tell how many more to that number.
Input/Output Table	This routine focuses on getting students to think about patterns.	Students have to fill in the number of the input/output table and in some cases create their own from scratch.
It Is/It Isn't	This routine can be used in a variety of ways. Students have to look at the topic and decide what it is and what it isn't. It is another way of looking at example, non-example.	Students discuss what something is and what it isn't.
Legs and Feet	Legs and feet is a great arithmetic routine which gets students to use various operations to figure out how many animals there could be by working with numbers.	Students look at different animals and think about how many legs and feet there could be given that number of animals.
Make Your Own Problem!	In this routine, students have to write their own problems.	Students have to write problems based on an equation or based on a topic.
Magic Square	In this fluency routine, students are working with math puzzles to figure out missing numbers.	There are a few different ways to do magic squares. One way is for students to figure out what the magic number is by calculating horizontally, vertically and diagonally.
Model That!	In this word problem routine, students are focusing on representing word problems in a variety of ways.	Students have to represent their thinking about a word problem with various models.
Missing Number	This routine focuses on students thinking about missing numbers.	Many of the missing number activities require that students reason about what number it should be.
Money Mix Up	This routine focuses on students' knowledge of money.	The money routines have students counting and comparing money quantities.
Number Bond It!	In this routine, students are working on decomposing numbers in a variety of ways.	Students use number bonds to break apart numbers in different ways.

Routine	Purpose	Description
Number of the Day	This routine focuses on students reasoning and modeling numbers in a variety of ways.	This activity has a given number and students have to represent that number in different ways.
Number Line It!	This routine focuses on sequencing numbers correctly.	Students have to put numbers in the correct sequence on the number path.
1-Minute Essay	In this routine, students have to think about, discuss and write about a concept.	Students write about a topic, their friend adds information and then they write some more about the topic.
Number Talk	This activity focuses on number sense. Students discuss solving different problems in different ways.	There are a few different ways that students do this activity. One of the ways is the teacher works with the students on solving a problem in a variety of ways. Another activity is that the teacher gives the students number strings around a specific concept for example subtracting 1 from a number and students work those problems and discuss the strategy.
Number Talk Puzzle	In this routine, students reason about numbers.	Students have to decide which numbers are missing to complete the problem.
Number Strings	In this routine, students are looking at the relationship among a set of problems.	Students work out the different problems and think about and discuss the various strategies they are using.
Open Array Puzzle	In this routine, students talk about open arrays.	Students have to decide which numbers are missing to make the open array true.
Patterns/Skip Counting	In this routine, students focus on patterns.	Many of the pattern activities require students to fill in a pattern and then make their own patterns.
Picture That!	In this routine, students discuss a picture.	Many of these activities require that students look at a picture and make up a math word problem about the picture.
Perimeter Problem	In this routine, students reason about the perimeter of shapes.	Students are given a specific perimeter and they have to come up with the possible length and width of the shape.

Routine	Purpose	Description
Place Value Puzzle	In this routine, students reason about numbers.	Students have to decide which numbers are missing to correctly complete the problem.
Reasoning Matrices	In this routine, students have to reason about a variety of things to figure out which ones go together.	Students are asked to reason about people given specific facts.
Rounding	In this routine, students work on rounding.	The rounding activities ask students to come up with the numbers that can be rounded rather than giving students a number to round.
Start With … Get to … By	In this routine, students focus on patterns and skip counting.	Students practice skip counting to different numbers.
Talk and Draw	In this routine, students have to talk and draw about the topic.	Students are given a topic and expected to discuss their thinking with math sketches.
10 Times as Much	In this routine, students have to reason about place value.	Students are given tasks to write numbers based on their place value.
True or False?	This routine focuses on students reasoning about what is true or false.	Students are given different things to think about like statements about shapes or equations and they have to state and prove whether they are true or false.
2 Arguments	In this reasoning routine, students are thinking about common errors that students make when doing various math tasks like missing numbers, working with properties and working with the equal sign.	Students listen to the way 2 different students approached a problem, decide who they agreed with and defend their thinking.
3 Truths and a Fib	In this routine, students have to consider whether or not statements are true or false.	Students are given 3 statements that are true and 1 statement that is false. They have to discuss why it is false.
Venn Diagram	In this routine, students create a Venn diagram with specific criteria.	Students are given a Venn diagram that they must fill out based on specific criteria.
Vocabulary Bingo	This is the traditional bingo game with a focus on vocabulary.	Students play vocabulary bingo but they have to discuss the vocabulary and make drawings or write definitions to show they know what the word means.

Routine	Purpose	Description
Vocabulary Brainstorm	In this routine, students have to brainstorm about vocabulary words.	Students have to write about different vocabulary words.
Vocabulary Fill-in	In this routine, the focus is on filling in sentences with the correct vocabulary word.	Students are given a list of sentences that have blanks and they have to fill in the blanks with words that make the sentences true.
Vocabulary Match	In this routine, the focus is on working with the math vocabulary from across the year.	Students match the vocabulary with the definition.
Vocabulary Talk	In this routine, the focus is on students talking about a topic using the correct vocabulary words.	Students talk about a topic and they have to write and draw about the vocabulary associated with that topic.
Vocabulary Tic Tac Toe	In this routine, students are working on math vocabulary words from across the year.	Students play tic tac toe by taking turns of choosing a square and then sketching or writing on the side to illustrate the word. Whoever gets three in a row first wins.
Why Is It Not?	This routine focuses on students looking at error patterns and correcting them.	Students have to look at error patterns and then discuss what the correct answer should be and prove why it should be that.
What Doesn't Belong?	This is a reasoning activity where students have to choose which objects they can group together and why. The emphasis is on justification.	Students have 4 squares. They have to figure out which object does not belong.
What's the Problem?	In this routine, students have to reason about numbers and contextualize them to real life situations.	Students are given the equation, expression or answer and they have to create a problem that matches it.
What's the Question? (3 Read Protocol)	The purpose is for students to slow down and consider all of the parts of the word problem.	Students have to read the problem 3 times. The first time they focus on the context. The second time they focus on the numbers. The third time they focus on asking questions that would make sense given the context.

Routine	Purpose	Description
What's the Story? Here's the Model	This routine focuses on students making sense of models and graphs.	Students have to look at the model and make up a story that matches it.
Why Is It Not?	In this routine, students are expected to reason about problems.	Students have to explain why the incorrect answer is not true.
Word Problem Fill-in	In this routine, students have to fill in numbers and make up and solve their own word problem.	Students fill in the blanks with numbers that they choose and then model and solve the word problem.

Questioning is the Key

To Unlock the Magic of Thinking, You Need Good Questions!

Figure 1.4

Launch Questions (Before the Activity)	Process Questions (During the Activity)
♦ What is this prompt asking us to do? ♦ How will you start? ♦ What are you thinking? ♦ Explain to your math partner, your understanding of the question. ♦ What will you do to solve this problem?	♦ What will you do first? ♦ How will you organize your thinking? ♦ What might you do to get started? ♦ What is your strategy? ♦ Why did you....? ♦ Why are you doing that? ♦ Is that working? Does it make sense? ♦ Is that a reasonable answer? ♦ Can you prove it? ♦ Are you sure about that answer? ♦ How do you know you are correct?
Debrief Questions (After the Activity) ♦ What did you do? ♦ How did you get your answer? ♦ How do you know it is correct? ♦ Can you prove it? ♦ Convince me that you have the correct answer. ♦ Is there another way to think about this problem?	**Partner Questions (Guide Student Conversations)** ♦ Tell me what you did. ♦ Tell me more about your model. ♦ Tell me more about your drawing. ♦ Tell me more about your calculations. ♦ Tell me more about your thinking. ♦ Can you prove it? ♦ How do you know you are right? ♦ I understand what you did. ♦ I don't understand what you did yet.

Daily Routines

Week 1 Teacher Notes

Monday: What Doesn't Belong?

When doing this routine, have the students do the calculations (in their journals, on scratch paper or on the activity page). Then, have them share their thinking with a friend. Then, pull them back to the group.

Tuesday: Vocabulary Match

This vocabulary match has many of the fourth grade words that students should know. Often when reviewing vocabulary it is good to review the grade level words mixed, meaning not by a specific category. Students should say the word and then find the matching definition. They should have some minutes to do this on their own and then an opportunity to go over their thinking with their math partner. Then, after about 5 minutes, come back together as a group and discuss the thinking. Ask students which words were tricky and which ones were easy. Also ask them were there any that they didn't recognize, that they have never seen before. Have them draw a little sketch by each word to help them remember the word.

Wednesday: Guess My Number

In this routine students are given some clues and they have to think about what the number could be given those clues.

Thursday: Number Talk

In this number talk you want the students to discuss their thinking with strategies and models. Ask students about the strategies that they might use.

Possible responses:

Break apart the thousands, hundreds, tens and ones. Do those calculations to get the partial sums. Then add everything back together.

Students might also change the 299 to 300 and then add it to 1279 and 6 (because they gave 1 to 299).

Friday: What's the Question? (3 Read Protocol)

The focus of today is to do a 3 read problem with the students. It is important to read the problem 3 times out loud as a choral read with the students.

First read: (Stop and visualize! What do you see?) What is this story about? Who is in it? What are they doing?

Second read: What are the numbers? What do they mean?

Third read: What are some possible questions we could ask about this story?

Possible Questions:

How far did Lucy run on Tuesday?

How far did she run altogether?

How much farther did she run on Wednesday than on Tuesday?

Which day did she run the farthest?

How far did she run the first 2 days?

Note: Focus on the vocabulary.

Focus on different types of comparative language so students get comfortable with words and phrases like: How many more? How much less? How many more to get the same amount as? How many fewer?

Week 1 Activities

Monday: What Doesn't Belong?

Look at the boxes. Pick the one that doesn't belong.

A.

$(2 \times 3) + (2 \times 3)$	$48 \div 4$
$108 \div 9$	$20 - 9$

B.

pints	liters
kilogram	gallons

Tuesday: Vocabulary Match

Match the word with the definition.

multiple	part of the answer to a multiplication problem
partial product	part of the answer to a division problem
partial quotient	a number you saw when you skip count by a number
dividend	a number in a multiplication problem
factor	the number you are dividing in a problem

Wednesday: Guess My Number

Read the clues and guess the number based on the clues.

I am greater than 57 and less than 85. I am a multiple of 7. The sum of my digits is 9. Who am I?	I am greater than $\frac{3}{4}$ and less than $\frac{7}{4}$. I am equivalent to $\frac{12}{8}$. Who am I?

Thursday: Number Talk

What are some ways to think about:

$$1279 + 299 + 7$$

Friday: What's the Question? (3 Read Protocol)

Read the problem 3 times. The first time talk about the situation. The second time talk about the numbers. The third time, ask questions based on the situation. Answer 2 of them. Discuss with your classmates.

Lucy ran $\frac{1}{5}$ of a mile on Monday, twice as far on Tuesday, and $\frac{4}{5}$ of a mile on Wednesday.

1)

2)

Week 2 Teacher Notes

Monday: Magic Square

This routine is great for practice. It is intriguing, fun and fast-paced. Students want to get the answer and they get right to work. In this first magic square students are trying to figure out the target number (the one that the digits add up to no matter which way you calculate them – horizontally, vertically or diagonally).

Have the students work on it on their own, then share their thinking with a partner. Then, bring everyone back to the whole group and have them discuss it.

Tuesday: Vocabulary Tic Tac Toe

These are quick partner energizers. Read all the words together. Then go! They have 7 minutes to play the game. They do rock, paper, scissors to start. They take turns choosing a word and explaining it to their partner. Then, they have to do a sketch or something to show they understand the word. Everybody should play the first game, if they have time they can play the next one. Note: It is important to call everyone back together at the end and talk about the vocabulary. Briefly go over the vocabulary.

Wednesday: Number Line It!

For this routine, students are expected to place the fractions on the number line in the correct order.

Thursday: Number Talk

For this routine, students are expected to say pairs of numbers that have a difference of 1,250, for example: 2,500 – 1,250 = 1,250. It is an open subtraction problem where there are many possible answers.

Friday: Elapsed Time

For this routine, students have to come up with possible times that fit the criteria. It is an open elapsed time problem, where there are many possibilities.

Week 2 Activities

Monday: Magic Square

Fill in the squares so that it makes a sum of $\frac{15}{12}$ in all directions, up, down, vertically, horizontally and diagonally.

	$\frac{9}{12}$	$\frac{2}{12}$
$\frac{3}{12}$	$\frac{5}{12}$	
		$\frac{6}{12}$

Tuesday: Vocabulary Tic Tac Toe

Play rock, paper, scissors to see who goes first. Then take turns, picking a square, saying what it means, drawing or writing something about the word on the side and then marking the word with an x or an o. Whoever gets 3 in a row first wins.

Game 1: Draw it or write the definition

quotient	dividend	difference
product	perimeter	tape diagram
multiple	area	open number line

Game 2: Draw it or write the definition

Draw a shape with 1 right angle.	Draw a parallelogram.	Draw a quadrilateral.
Draw a rectangular prism.	Draw a hexagon that does not look like this. ⬢	Draw a rhombus.
Draw a shape with 1 pair of parallel sides.	Draw some examples of polygons.	Draw a kite.

Wednesday: Number Line It!

Place these numbers on the number line in order from least to greatest. Be as exact as possible.

$\frac{4}{12}$ $\frac{2}{6}$ $\frac{3}{5}$ $\frac{1}{2}$

Compare these with your partner. Explain your thinking. How do you know you are correct?

Thursday: Number Talk

Discuss pairs of numbers that have a difference of 1,250.

Friday: Elapsed Time

Read the problem and discuss it with your partner. Be ready to talk about it in the group.

Sue left her house and came back 45 minutes later. What time could she have left? What time could she have come back?

Name 3 possibilities:

1.

2.

3.

Week 3 Teacher Notes

Monday: Draw and Talk

Students will draw and talk about the various angles.

Tuesday: Frayer Model

Students use the word to fill in the diagram.

Wednesday: Number of the Day

This routine is important and reviews the place value skills. There are both closed and open items in the routine. Give the students about 5 minutes to work on this and then discuss it with their math partner. Then, come back together as a class and talk about what students did.

Thursday: Number Strings

Students should discuss the string as a whole class. You want the emphasis to be on the relationships between the numbers. Students should be thinking about if they know one of the base facts how it helps them with the other facts. Sometimes, people call these "helper" facts.

Friday: Model That!

Students should model the problem in more than one way.

Week 3 Activities

Monday: Draw and Talk

Draw an acute, obtuse and a right angle. Explain each drawing.

Tuesday: Frayer Model

Fill in the blanks based on the word.

Perimeter

Formula	Example
Picture	Non-example

Wednesday: Number of the Day

Fill in the boxes based on the number of the day.

$$\frac{9}{8}$$

Word form	___ + ___ = ____ ___ − ___ = ____
____ is greater than ____ ___ is less than ______ _____ is equivalent to ____	Decompose

Thursday: Number Strings

Talk about these numbers with your math partner and then the whole class.

$\frac{1}{10}+\frac{1}{10}$

$\frac{1}{10}+\frac{2}{100}$

$\frac{1}{10}+\frac{3}{10}$

$\frac{1}{10}+\frac{5}{100}$

Friday: Model That!

Model this problem in 2 ways.

$3 \times \frac{3}{4}$

A. Number line

B. Another way

Week 4 Teacher Notes

Monday: Convince Me!

Students have to defend their thinking with numbers, words and pictures. The focus is on the discussion that students have with their math partners and the whole group.

This is true! I can prove it with …
This is the same because …
I am going to use __________ to show my thinking.
I am going to defend my answer by __________.

Tuesday: Frayer Model

Students fill in the diagram based on the word of the day.

Wednesday: Greater than, Less Than, in Between

In this routine, students are thinking about the number relationships. Have the students do it on their own and then talk with their math partners. Then, bring it back together as a class and discuss it.

Thursday: Number Talk

This is a typical number talk where students are thinking about the ways in which they can solve this subtraction problem. You want students to think about partial differences, counting up, and compensation. For example:

One way to do this is to have the students adjust the numbers by adding 3 to each one. The new problem is 1,537 – 1,400 which is a much easier problem.

Friday: Picture That!

Students look at the picture and tell and solve a story about it.

Week 4 Activities

Monday: Convince Me!

Discuss this problem with your neighbor. Convince them that it is true using numbers, words and models.

$$2 \times 7 \times 5 = 10 \times 7$$

Tuesday: Frayer Model

Fill in the boxes based on the word.

Remainder

Definition	Examples
Give a picture example	**Non-examples**

Wednesday: Greater Than, Less Than, in Between

Read the prompts and fill in the boxes.

$\frac{2}{3}$ $\frac{1}{2}$ 1

Name a fraction greater than $\frac{1}{2}$	Name a fraction less than $\frac{2}{3}$	Name a fraction less than 1
Name a fraction greater than $\frac{1}{2}$	Name a fraction in-between $\frac{1}{2}$ and 1	Name a fraction in-between $\frac{2}{3}$ and 1

Thursday: Number Talk

Discuss this problem with a partner and your class.

What are some ways to subtract 1,534 – 1,397?

Friday: Picture That!

Tell a fraction story about these animals.

Story:

Equation:

Week 5 Teacher Notes

Monday: 2 Arguments

Students look at this problem and discuss it with their math partner. The emphasis should be on proving their thinking. Discuss how what they said is true. Then, everyone will come back together and discuss it.

Tuesday: Talk and Draw

Students are going to discuss, write and draw about division words.

Wednesday: Start With … Get to … By

Students have to start with a number, get to a target number by counting in a specific way.

Thursday: British Number Talk

Give students about 5 minutes to work on their own or with partners to come up with some problems, hopefully from each category. This is so that they stretch. You don't want them to only choose the easy problem. Then, students should share what they did with class.

Friday: Make Your Own Problem!

Give students about 5 minutes to do their own fill in problems and share with their math partner. Their partner has to tell them if their problem makes sense. Then, some students will share out their thinking with the class.

Week 5 Activities

Monday: 2 Arguments

Look at the problem. Discuss with a neighbor. Be ready to talk with the whole class.

$240 = 40 \times ?$

John said the answer was 280.

Maria said the answer was 6.

Who do you agree with?

Why?

Tuesday: Talk and Draw

Write about all the words you know that have to do with division. Draw examples of the words. Discuss with your math partner.

Wednesday: Start With … Get to … By

Start with $\frac{1}{4}$ get close to 4 skip counting by $\frac{3}{4}$.

Start with $\frac{5}{8}$ get a little bit more than 2 by skip counting by $\frac{1}{4}$.

Start with $\frac{3}{5}$ get to 2 skip counting by $\frac{2}{10}$.

Start with $\frac{1}{2}$ get to 3 skip counting by $\frac{1}{2}$.

Thursday: British Number Talk

Pick a number from each circle. Subtract. Write the expression under the way you solved it. For example, 399 – 301. I can do that in my head.

I can do it in my head.	I can do it with a model.	I can do it using a written strategy or algorithm.

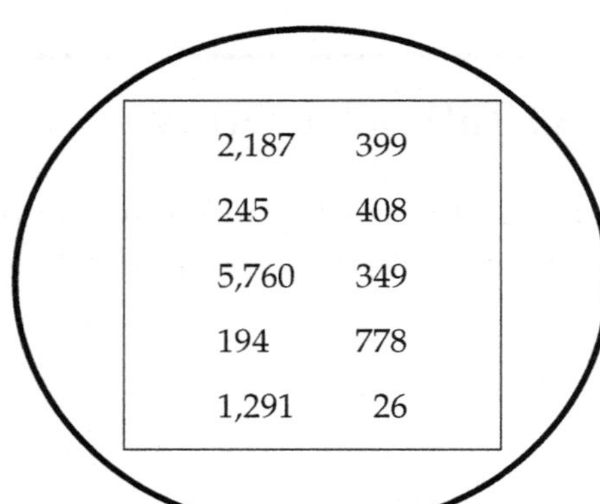

Friday: Make Your Own Problem!

Fill in the blanks.

Ray had _______marbles. His brother had ___ times as many as he did. How many did his brother have? How many did they have altogether?

Model it!

Number sentence (equation): ____________________

Week 6 Teacher Notes

Monday: It Is/It Isn't

In this routine you want students to be focusing on the vocabulary. Encourage students to use the word bank. This is a scaffold only though, to get them started. The conversation might sound something like this: It is an even number. It is a 4-digit number. It is not odd. It is a multiple of 5.

Tuesday: 1-Minute Essay

Give students the designated part of the time to write and then share and then write again and then share out with the class.

Wednesday: Find and Fix the Error

Students have to look at the problem with their math partner and discuss what is incorrect. Then, they figure out how they would fix it. They should be ready to come back and discuss it with the class.

Thursday: Number Strings

Students should discuss the string as a whole class. You want the emphasis to be on the relationships between the numbers. In this string we are working on what happens when you multiply by multiples of ten. You want students to talk about place value.

Friday: Fill in the Problem!

Students pick a number and solve the problem. The focus here is that they have to also make a tape diagram that models their problem.

Week 6 Activities

Monday: It Is/ It Isn't

Discuss what this number is and isn't. Use your math vocabulary.

1,080

It Is	It Isn't

Word bank: multiples, factor, odd, even, composite, prime, square, divisible by.

Tuesday: 1-Minute Essay

(For 30 seconds) Write everything you can about equivalent fractions. Use numbers, words, and pictures.

(15 seconds) Now switch with a neighbor and add 1 thing to their list.

(15 seconds) Now add 1 more thing to your list.

Wednesday: Find and Fix the Error

John did this. The answer is wrong. Find and fix the error.

$$\begin{array}{r} 1{,}000 \\ -\ 178 \\ \hline 1{,}178 \end{array}$$

1. What is wrong?
2. Why can't you do what John did?
3. Fix it.
4. Explain your thinking to your partner and then the whole group.

Thursday: Number Strings

Discuss these problems with your neighbor. Be prepared to discuss these problems with the whole class.

0.5×10

0.5×100

$0.5 \times 1{,}000$

$0.5 \times 10{,}000$

Friday: Fill in the Problem!

Pick a number. Solve. Discuss your thinking with your partner. Be ready to talk with the class.

In the aquarium there were 10 fish. This was _______ (2 or 5) times as many turtles as fish. How many animals were there altogether?

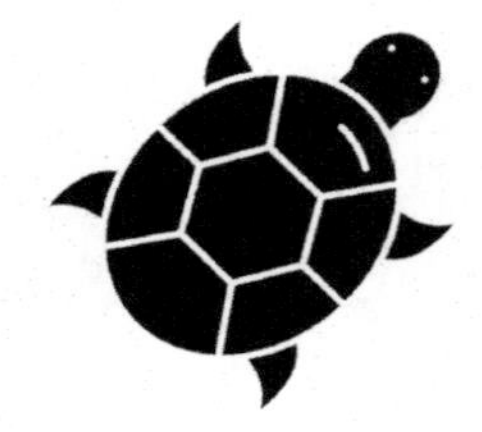

Model it with a tape diagram.

Week 7 Teacher Notes

Monday: 3 Truths and a Fib

This is a place value reasoning activity. Students have to compare decimals thinking about tenths, hundredths and equality.

Tuesday: Vocabulary Brainstorm

Students have to think and then write about the word in all of the bubbles. They first share their thinking with a partner and then share their thinking with the class.

Wednesday: Input/Output Table

Students have to fill in the input/output table.

Thursday: Number Talk

In this number talk you want students to either count up or think about adjusting the numbers to make them easier. Students could make an easier problem such as $2.00 + $3.53.

Friday: Get Close to

Students have to pick 2 two-digit numbers that have a product that is close to 1,000. This is tricky. Students struggle with doing it in the beginning. After doing it a few times it becomes easier to do.

Week 7 Activities

Monday: 3 Truths and a Fib

Which one is false? Why? Explain to your neighbor and then the group.

10.4 > 10.04	7.10 < 7.01
9.20 > 9.02	10.40 = 10.4

Tuesday: Vocabulary Brainstorm

In each thought cloud write or draw something that has to do with multiplication.

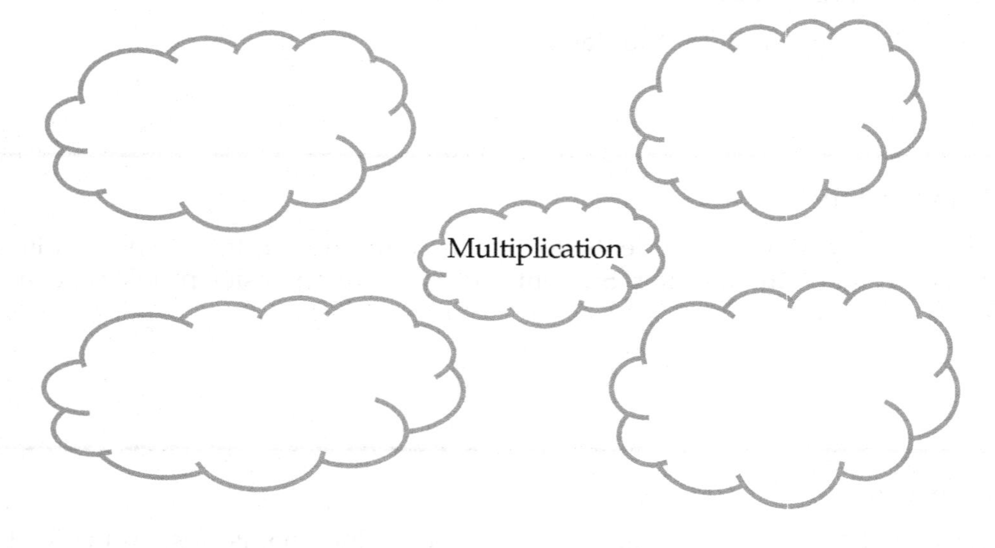

Wednesday: Input/Output Table

Create an input-output table where the rule is to multiply by 12.

In	Out

Thursday: Number Talk

Discuss this with your math partner. Be prepared to discuss this with the whole class.

What are some ways to solve:

$1.99 + $3.54

Friday: Get Close to

Multiply 2 two-digit numbers that have a product that is close to 1,000.

Week 8 Teacher Notes

Monday: Alike and Different

How are .75 and .80 alike and different?

Students should use terms like greater than, less than, multiples, factors, divisible by…

Tuesday: Vocabulary Tic Tac Toe

These are quick partner energizers. Read all the words together. Then go! They have 7 minutes to play the game. They do rock, paper, scissors to start. They take turns choosing a word and explaining it to their partner. Then, they have to do a sketch or something to show they understand the word. Everybody should play the first game, if they have time, they can play the next one.

Note: It is important to call everyone back together at the end and talk about the vocabulary. Briefly go over the vocabulary, this is all vocabulary that they should know.

Wednesday: Number Bond It!

Number bonds are important because they build flexibility. Students should be thinking about how to compose and decompose numbers in a variety of ways. Have the students work on this by themselves first and then discuss with a partner and finally with the entire class.

Thursday: British Number Talk

Give students about 5 minutes to work on their own or with partners to come up with some problems, hopefully from each category. This is so that they stretch. You don't want them to only choose the easy problem. Then, students should share what they did with class

Friday: What's the Story? Here's the Model

The students have to write a division story where the remainder is 2 cookies.

Week 8 Activities

Monday: Alike and Different

How are .75 and .80 alike? How are they different?

Word bank: multiples, factors, skip counting, divisible by, digits, place value, tenths, hundredths, ones.

Tuesday: Vocabulary Tic Tac Toe

Play rock, paper, scissors to see who goes first. Then take turns, picking a square, saying what it means, drawing or writing something about the word on the side and then marking the word with an x or an o. Whoever gets 3 in a row first wins.

ml	mm	m
kg	mg	km
cm	kl	l

area	perimeter	yard
feet	inch	oz
gallon	quart	pint

Wednesday: Number Bond It!

Show how to break apart $1\frac{5}{6}$ in 3 different ways!

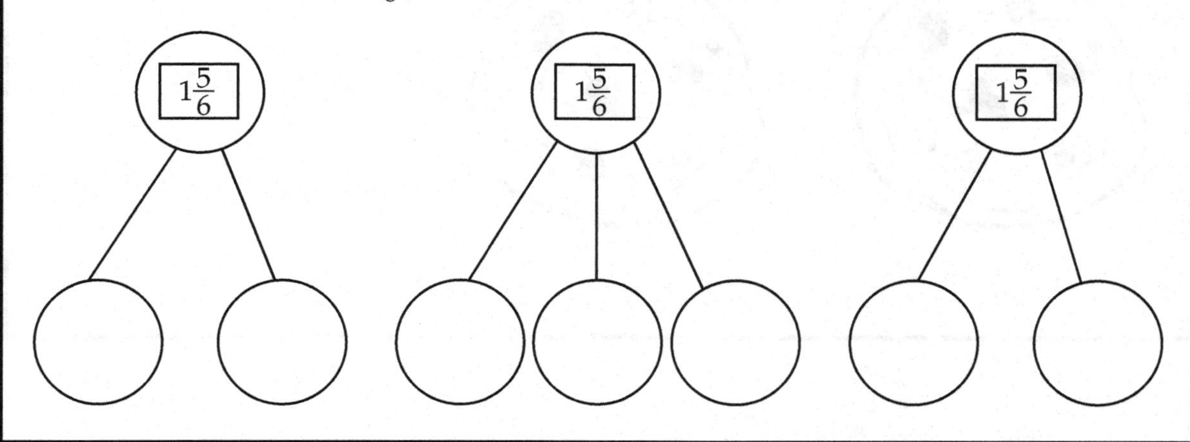

Thursday: British Number Talk

Pick a number from each circle. Multiply. Write the expression under the way you solved it. For example, 10×11 is 110. I can do that in my head.

I can do it in my head.	I can do it with a model.	I can do it using a written strategy or algorithm.

A.

12 22 13 14

15 26

71 82 99 100

110 10

B.

0 11 21 13 41 15

16 17 28 39 10 11

12 24 33 44 50 62

75 83 99

Friday: What's the Story? Here's the Model

The remainder is 2 cookies. It was a division story. What was the story?

Week 9 Teacher Notes

Monday: Number Line It!

Students have to read the problem and answer the question. It is an open question so that there could be various answers.

Tuesday: Vocabulary Bingo

Students play bingo with the math vocabulary. The point here is to be able to discuss these words and talk about what they mean and how we use them in math.

Wednesday: Fraction of the Day

This activity provides ongoing practice with fractions. Students use the fraction to practice the different items in the boxes.

Thursday: Number Talk Puzzle

Students have to talk about the missing numbers.

Friday: Graph Talk

Students are given a graph. They have to make up a story and tell it to their partner and then be prepared to discuss it with the entire class.

Week 9 Activities

Monday: Number Line It!

Write 3 fractions that can be in between the ones on this number line.

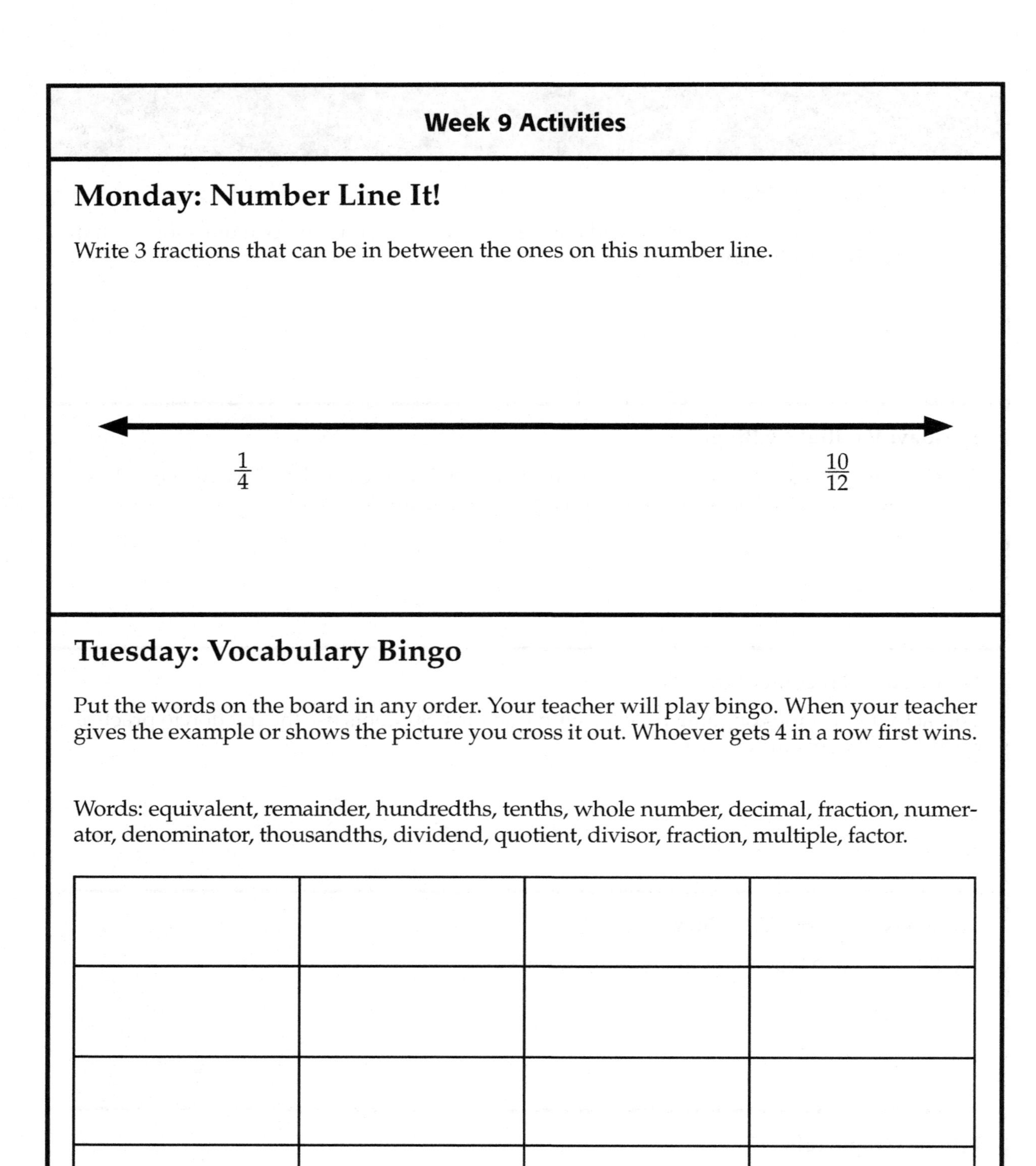

$\frac{1}{4}$ $\frac{10}{12}$

Tuesday: Vocabulary Bingo

Put the words on the board in any order. Your teacher will play bingo. When your teacher gives the example or shows the picture you cross it out. Whoever gets 4 in a row first wins.

Words: equivalent, remainder, hundredths, tenths, whole number, decimal, fraction, numerator, denominator, thousandths, dividend, quotient, divisor, fraction, multiple, factor.

Wednesday: Fraction of the Day

Fill in the boxes based on the fraction.

$$\frac{5}{4}$$

Word form:	How far from 1?	Draw a model.
____ is greater than ____ ____ is less than ____ ____ is equivalent to _____ Add a fraction that is greater than this number. Add a fraction that is less than this number. Plot an equivalent fraction on the number line.		$\frac{5}{4} \times \frac{1}{2} =$ $\frac{5}{4} - \_\_\_ = \frac{1}{16}$ $\frac{5}{4} + \frac{2}{8} =$

Thursday: Number Talk Puzzle

What are the missing numbers? Fill in numbers that make the puzzle true.

$$\begin{array}{r} 2.\square 7 \\ + \square.5\square \\ \hline 8.56 \end{array}$$

Friday: Graph Talk

Here's the graph.	What's the story? Make up a story and tell it to your partner. Explain your thinking. Be ready to share your thinking to the whole group.

				x
				x
	x	x		x
x	x	x		x
x	x	x	x	x
x	x	x	x	x
x	x	x	x	x
$\frac{1}{2}$	1	$1\frac{1}{2}$	2	$2\frac{1}{2}$

Week 10 Teacher Notes

Monday: Reasoning Matrices

Students have to fill in the missing number. Watch for common errors such as multiplying the 2 given numbers or subtracting them. Focus the conversation on reasoning about the equation. Focus on the equal sign as having the meaning "is the same as."

Tuesday: 1-Minute Essay

Students have to write about the word using numbers, words and pictures. They then read each other's work and add to it and then write some more.

Wednesday: Decimal of the Day

Decimal of the day is important and reviews the place value skills. There are both closed and open items in the routine. Give the students about 5 minutes to work on this and then discuss it with their math partner. Then, come back together as a class and talk about what students did.

Thursday: British Number Talk

Give students about 5 minutes to work on their own or with partners to come up with some problems, hopefully from each category. This is so they stretch. You don't want them to only choose the easy problem. Then, students should share what they did with class.

Friday: What's the Story? Here's the Model

Students have to tell a story about the bar diagram. This is a reasoning activity. We want students to be able to contextualize numbers.

Week 10 Activities

Monday: Reasoning Matrices

Fill in the missing numbers.

$5 \times \_\_\_ = 90$

$630 = \_\_\_ \times 9$

$20 = 100 \div ?$

$\_\_\_ \times \_\_\_\_ = 540$

Tuesday: 1-Minute Essay

(For 30 seconds) Write everything you can about place value. Use numbers, words, and pictures.

(15 seconds) Now switch with a neighbor and add 1 thing to their list.

(15 seconds) Now add 1 more thing to your list.

Wednesday: Decimal of the Day

Fill in the boxes based on the decimal of the day.

.92

Word form	.10 more .10 less	Round to nearest tenth ____
Expanded form	____ + ____	_____ − _____ =

Plot the number, a number close to the number, and a number far from the number.

Thursday: British Number Talk

Pick a number from each circle. Divide. Write the expression under the way you solved it. For example, 4,809 ÷ 9. I can do it with a model.

I can do it in my head.	I can do it with a model.	I can do it using a written strategy or algorithm.

A.

284	110	
185	243	
120	127	
402	36	4,809
140	216	1,450

B.

	10	12	30
4	25	16	
7	18	9	
52	2	41	

Friday: What's the Story? Here's the Model

Look at the bar diagram below. Write a story that matches the bar diagram.

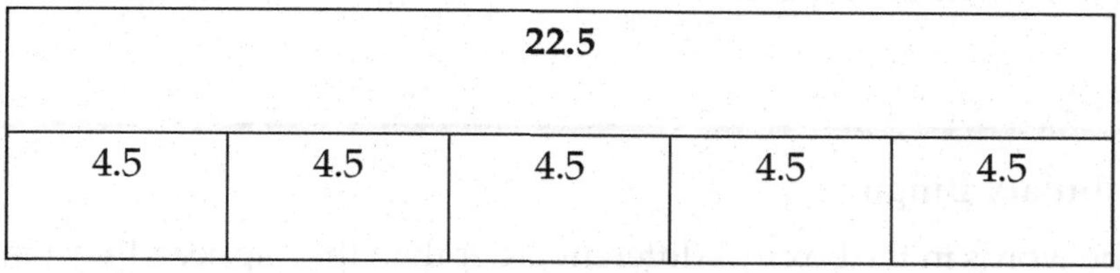

22.5				
4.5	4.5	4.5	4.5	4.5

Story:

Equation:

Week 11 Teacher Notes

Monday: Find and Fix the Error

Students have to look at the problem and figure out what is wrong and fix it.

Tuesday: Vocabulary Bingo

Students put the words in the box in a different order than they appear. Play bingo. When the teacher calls a word, cover it. Whoever gets 4 in a row vertically, horizontally or 4 corners wins.

Wednesday: Why Is It Not?

Students look at each problem and think about why the given answer is not correct. They have to discuss it and then explain their thinking to the class.

Thursday: British Number Talk

Give students about 5 minutes to work on their own or with partners to come up with some problems, hopefully from each category. This is so they stretch. You don't want them to only choose the easy problem. Then, students should share what they did with class.

Friday: What's the Question? (3 Read Protocol)

The focus of today is to do a 3 read problem with the students. It is important to read the problem 3 times out loud as a choral read with the students.

First read: (Stop and visualize! What do you see?) What is this story about? Who is in it? What are they doing?

Second read: What are the numbers? What do they mean?

Third read: What are some possible questions we could ask about this story? After the students read it, write down, make up questions and then solve them. They should discuss their answers with the class.

Possible Questions:

Who ate more, Mike or Tom?

Who ate more out of all 3 of them?

How much did Mike and Tom eat altogether?

Note: Focus on the vocabulary.

Focus on different types of comparative language so students get comfortable with words and phrases like: How many more? How much less? How many more to get the same amount as? How many fewer?

Week 11 Activities

Monday: Find and Fix the Error

Sara solved the problem this way:

$\frac{2}{5} + \frac{3}{5} = \frac{5}{10}$

Why is it wrong?

What did she do wrong?

How can we fix it?

Tuesday: Vocabulary Bingo

Put the words in the box in a different order than they appear. Play bingo. When the teacher calls a word, cover it. Whoever gets 4 in a row vertically, horizontally or 4 corners wins.

Words: mixed number, fractions, dividend, divisor, factor, multiple, addend, partial product, partial quotient, partial sums, partial differences, equal, decimal, multiply, even, odd.

Write a word in each space.

Wednesday: Why Is It Not?

Think about it. Talk about it with a partner. Share with the class.

A. 10 = _____ + 5.6 Why is it not 15.6?	B. Why is this not a parallelogram? 

Thursday: British Number Talk

Pick a number from each circle. Add. Write the expression under the way you solved it. For example, $\frac{1}{2} + \frac{1}{2}$. I can do it in my head.

I can do it in my head.	I can do it with a model.	I can do it using a written strategy or algorithm.

A.

$\frac{1}{2}$ $\frac{3}{4}$ $\frac{2}{4}$ $\frac{1}{3}$
$\frac{2}{3}$ $\frac{3}{3}$ $\frac{1}{5}$ $\frac{2}{5}$ $\frac{3}{5}$ $\frac{4}{5}$ $\frac{1}{6}$
$\frac{2}{6}$ $\frac{3}{6}$ $\frac{4}{6}$ $\frac{5}{6}$ $\frac{1}{8}$ $\frac{2}{8}$
$\frac{3}{8}$ $\frac{4}{5}$ $\frac{5}{8}$ $\frac{6}{8}$ $\frac{7}{8}$ $\frac{1}{10}$
$\frac{2}{10}$ $\frac{3}{10}$ $\frac{4}{10}$ $\frac{5}{10}$ $\frac{6}{10}$
$\frac{7}{10}$ $\frac{8}{10}$ $\frac{9}{10}$

B.

$\frac{1}{2}$ $\frac{3}{4}$ $\frac{2}{4}$ $\frac{1}{3}$
$\frac{2}{3}$ $\frac{3}{3}$ $\frac{1}{5}$ $\frac{2}{5}$ $\frac{3}{5}$ $\frac{4}{5}$
$\frac{1}{6}$ $\frac{2}{6}$ $\frac{3}{6}$ $\frac{4}{6}$ $\frac{5}{6}$ $\frac{1}{8}$
$\frac{2}{8}$ $\frac{3}{8}$ $\frac{4}{5}$ $\frac{5}{8}$ $\frac{6}{8}$ $\frac{7}{8}$
$\frac{1}{10}$ $\frac{2}{10}$ $\frac{3}{10}$ $\frac{4}{10}$ $\frac{5}{10}$ $\frac{6}{10}$
$\frac{7}{10}$ $\frac{8}{10}$ $\frac{9}{10}$

Friday: What's the Question? (3 Read Protocol)

Read the problem 3 times. The first time talk about the situation. The second time talk about the numbers. The third time, ask questions based on the situation. Answer 2 of them.

Mike and his brothers shared a candy bar. Mike ate $\frac{2}{6}$ of it. Tom ate $\frac{3}{12}$ of it. Joe ate the rest.	What are some questions that you can ask?

Week 12 Teacher Notes

Monday: Legs and Feet

These are reasoning problems. Students look at the first picture and then reason through the next problems using the first picture as a launchpad.

Tuesday: Vocabulary Brainstorm

In each thought cloud students write or draw something that has to do with decimals.

Wednesday: 3 Truths and a Fib

Students have to read the statements and decide which 3 are true and which one is not. They then discuss with their neighbor which one is false. They must explain their thinking to their neighbor and then to the class.

Thursday: Number Talk

Students discuss the problem and talk about ways to solve it.

Friday: What's the Problem?

Students have to write a word problem where $\frac{4}{6}$ of a cup of butter is the answer. They should also model it and write the equation.

Week 12 Activities

Monday: Legs and Feet

There is a chicken and a cow.

A. How many legs?	B. If there are 18 legs and there has to be a chicken, a cow, and a cricket, how many animals and what types are there?
C. If there are 32 legs and there has to be a chicken and a cow, how many animals and what types are there?	D. Make up your own problem.

Tuesday: Vocabulary Brainstorm

In each thought cloud, write or draw something that has to do with decimals.

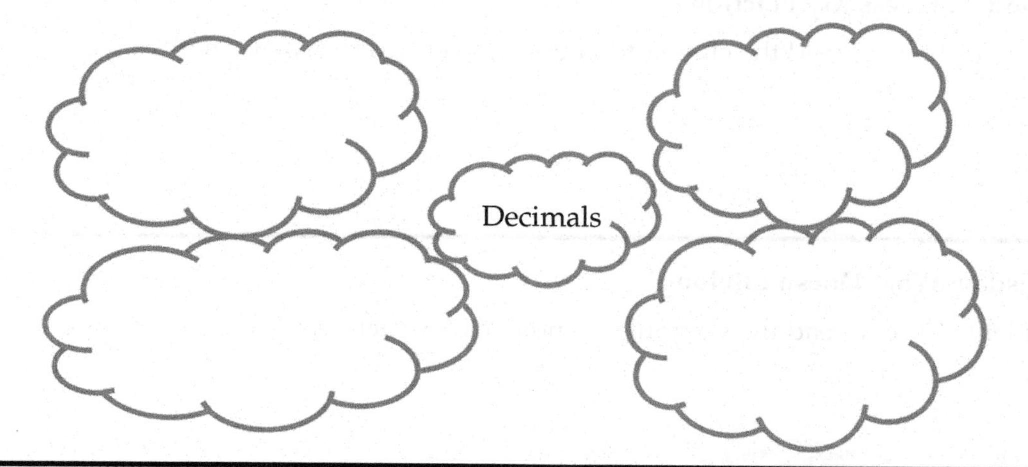

Wednesday: 3 Truths and a Fib

$\frac{1}{2} = \frac{4}{8}$	$\frac{2}{4} = \frac{5}{8}$
$1\frac{1}{2} = \frac{3}{2}$	$\frac{2}{2} = \frac{8}{8}$

Thursday: Number Talk

What are some ways to subtract \$4.21 – \$3.99?

Friday: What's the Problem?

Write a word problem where this is the answer.

The answer is $\frac{4}{6}$ of a cup of butter.

Word problem:

Model:

Equation:

Week 13 Teacher Notes

Monday: Guess My Fraction

Students have to read the clues and guess which fraction is being named.

Tuesday: What Doesn't Belong?

Students have to read the vocabulary and decide which word does not belong.

Wednesday: Fraction of the Day

Students fill in the boxes based on the given fraction.

Thursday: Number Talk

Students discuss ways to solve the expression.

Friday: What's the Story? Here's the Model

In this problem students have to look at the tape diagram and then tell a story that matches it.

Week 13 Activities

Monday: Guess My Fraction

Read the clues and then guess which fraction is being named.

I am a fraction that is greater than $\frac{1}{2}$. I am less than $\frac{3}{4}$. Who am I?	$\frac{5}{12}$ $\frac{4}{8}$ $\frac{7}{12}$ $\frac{3}{8}$

Tuesday: What Doesn't Belong?

Read the vocabulary and decide which word does not belong.

A.

quotient	remainder
addend	dividend

B.

Distance around the outside edge of a figure	$2l + 2w$
perimeter	area

Wednesday: Fraction of the Day

Fill in the boxes based on the given fraction.

$$1\frac{2}{3}$$

Word form	Picture form
Plot it on a number line	$\_\_\_ + \_\_\_ = 1\frac{2}{3}$ $\_\_\_ - \_\_\_ = 1\frac{2}{3}$ $1\frac{2}{3} \times \_\_\_ = \_\_\_$

Thursday: Number Talk

What are some ways to think about and show:

$$35 \times 71$$

Friday: What's the Story? Here's the Model

Look at the tape diagram. Think of a story and discuss it with your partner. Be ready to share out with the whole group.

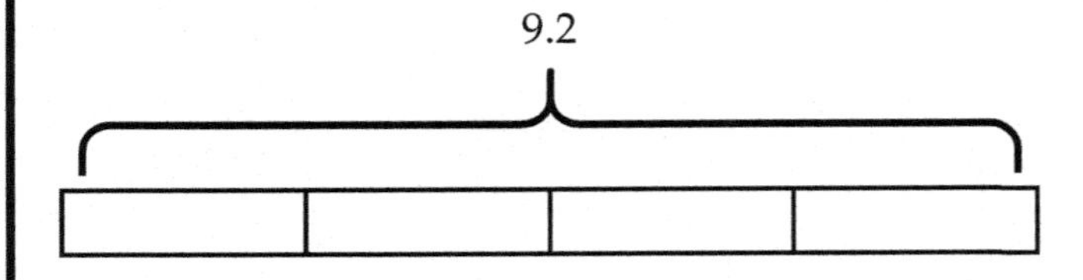

Week 14 Teacher Notes

Monday: Always, Sometimes, Never

Students have to decide if the statements are always, sometimes or never true.

Tuesday: Alike and Different

Students discuss how the two terms are alike and different.

Wednesday: Venn Diagram

Students have to follow the instructions, make a Venn diagram and then discuss it.

Thursday: British Number Talk

Give students about 5 minutes to work on their own or with partners to come up with some problems, hopefully from each category. This is so they stretch. You don't want them to only choose the easy problem. Then, students should share what they did with class.

Friday: Perimeter Problem

Students have to reason about the problem and give possible solutions.

Week 14 Activities

Monday: Always, Sometimes, Never

Read each statement. Decide if it is always, sometimes, or never true. Discuss your thinking with your math partner and the whole group.

A.	B.
$\frac{1}{4}$ is less than $\frac{1}{2}$.	The decimal with the most digits is the larger number.

Tuesday: Alike and Different

How are these alike and how are they different? Share your thinking with your neighbor. Be ready to share with the whole class.

decimals and fractions

Wednesday: Venn Diagram

Draw a Venn diagram and fill it in according to the instructions.

Circle A: Factors of 36

Circle B: Factors of 48

Overlap: Common factors of 36 and 48

Thursday: British Number Talk

Pick a number from each circle. Make a fraction subtraction problem. Write the problem under the way you solved it.

I can do it in my head.	I can do it with a model.	I can do it using a written strategy or algorithm.

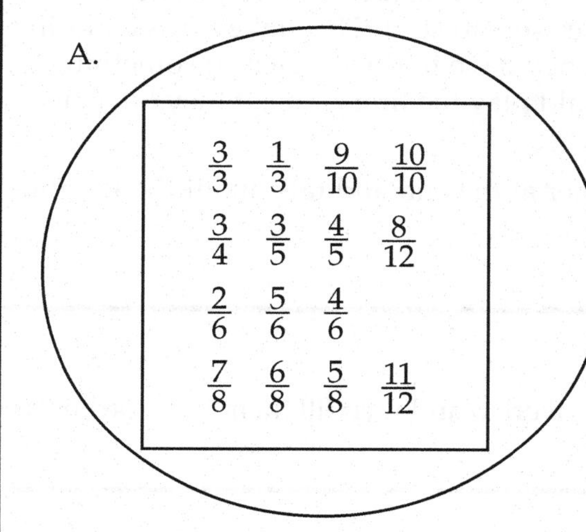

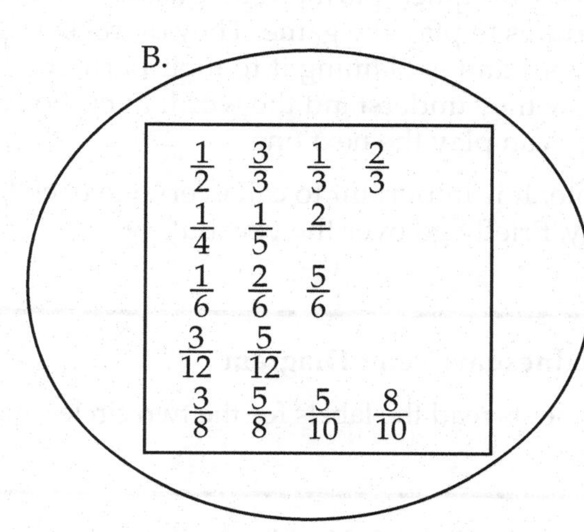

Friday: Perimeter Problem

There was a fence that had an area of 120 square feet. What are the possible length and widths of the fence?

Week 15 Teacher Notes

Monday: Number Line It!

Students need to place the numbers on the number line in the correct order. Students should try to be as exact as possible.

Tuesday: Vocabulary Tic Tac Toe

These are quick partner energizers. Read all the words together. Then go! They have 7 minutes to play the game. They do rock, paper, scissors to start. They take turns choosing a word and explaining it to their partner. Then, they have to do a sketch or something to show they understand the word. Everybody should play the first game, if they have time, they can play the next one.

Note: It is important to call everyone back together at the end and talk about the vocabulary. Briefly go over the vocabulary.

Wednesday: Venn Diagram

Students read the labels for the two circles and the overlap and then fill them in accordingly.

Thursday: British Number Talk

In this number talk you want the students to discuss their thinking with strategies and models. Ask students about the strategies that they might use.

Friday: What's the Question? (3 Read Protocol)

The focus of today is to do a 3 read problem with the students. It is important to read the problem 3 times out loud as a choral read with the students.

First read: (Stop and visualize! What do you see?) What is this story about? Who is in it? What are they doing?

Second read: What are the numbers? What do they mean?

Third read: What are some possible questions we could ask about this story? After the students read it, write down, make up questions and then solve them. They should discuss their answers with the class.

Possible Questions:

How much more did Lucy have than her brother?

Who had more?

Note: Focus on the vocabulary.

Focus on different types of comparative language so students get comfortable with words and phrases like: How many more? How much less? How many more to get the same amount as? How many fewer?

Week 15 Activities

Monday: Number Line It!

Place these fractions on the number line from least to greatest.

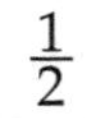

$\frac{1}{2}$ $\frac{3}{12}$ $\frac{6}{8}$ $\frac{7}{2}$ $\frac{9}{12}$ 1

Tuesday: Vocabulary Tic Tac Toe

Read all the words together. Then go! They have 7 minutes to play the game. Do rock, paper, scissors to start. Then, take turns choosing a word and explaining it to your partner. Then, do a sketch or something to show that you understand the word. Everybody should play the first game, if they have time, they can play the next one.

Name the fraction representation or draw it.

Mixed number	denominator	
	$\frac{5}{5}$	Parts of the whole
equivalent	$\frac{9}{8}$	

Say what the abbreviation stands for. Give an example of something that is measured in those units.

cm	m	l
In.	k	mm
ml	ft	kg

Wednesday: Venn Diagram

Look at the pictures. Fill in the Venn diagram.

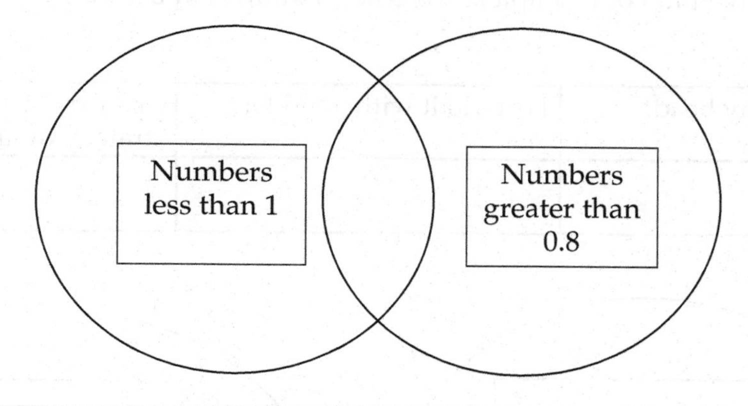

$\frac{1}{2}$	$1\frac{1}{2}$	$\frac{9}{10}$	$\frac{8}{100}$	0.9	0.09	1.8	0.88

Thursday: British Number Talk

Pick a number from each circle. Make a multiplication problem. Write the problem under the way you solved it. For example, 2 x $1\frac{1}{2}$ is 3. I know that because $1\frac{1}{2}$ + $1\frac{1}{2}$ is 3. I did it in my head.

I can do it in my head.	I can do it with a model.	I can do it using a written strategy or algorithm.

A.

2 3 5 9

8 6

7 4

B.

$1\frac{1}{2}$ $3\frac{1}{4}$ $6\frac{1}{2}$

$2\frac{3}{4}$ $4\frac{1}{2}$

$\frac{1}{3}$ $\frac{1}{2}$ $\frac{1}{4}$

Friday: What's the Question? (3 Read Protocol)

Read the problem 3 times. The first time talk about the situation. The second time talk about the numbers. The third time, ask questions based on the situation. Answer 2 of them.

Lucy had $.85. Her brother had $.48. Her sister had $1.33.

Week 16 Teacher Notes

Monday: Patterns/Skip Counting

Students have to write the pattern.

Tuesday: It Is/It Isn't

Students use the word bank to discuss the number with the frame.

Wednesday: Convince Me!

Students have to defend their thinking with numbers, words and pictures. The focus is on the discussion that students have with their math partners and the whole group.

This is true! I can prove it with …
This is the same because …
I am going to use __________ to show my thinking.
I am going to defend my answer by __________.

Thursday: British Number Talk

Give students about 5 minutes to work on their own or with partners to come up with some problems, hopefully from each category. This is so they stretch. You don't want them to only choose the easy problem. Then, students should share what they did with class.

Friday: Make Your Own Problem!

Students have to write a word problem using either fractions or decimals.

Week 16 Activities

Monday: Patterns/Skip Counting

Write a pattern that adds 7 and subtracts 2. Write 5 terms in your pattern.

Tuesday: It Is/It Isn't

Discuss the number using the frame it is/it isn't.

.05

It Is	It Isn't

Word bank: hundredths, tenths, thousandths, whole number, multiple, factor.

Wednesday: Convince Me!

$$150 \times 9 = (100 \times 9) + (50 \times 9)$$

Use numbers, words and pictures.

Thursday: British Number Talk

Pick a number from each circle. Multiply. Write the expression under the way you solved it. For example, $6 \times \frac{3}{10}$. I can do that with a model.

I can do it in my head.	I can do it with a model.	I can do it using a written strategy or algorithm.

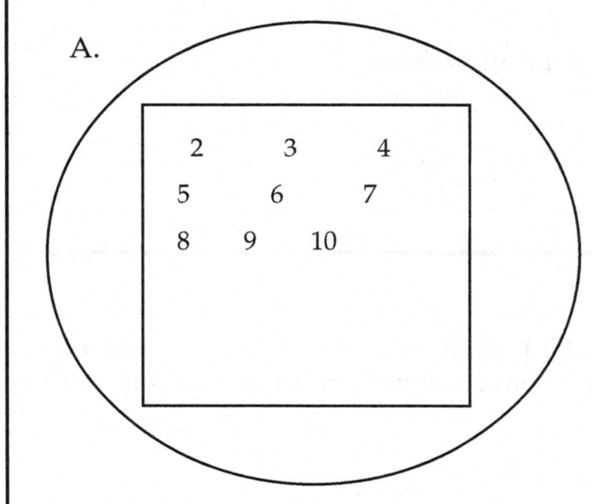

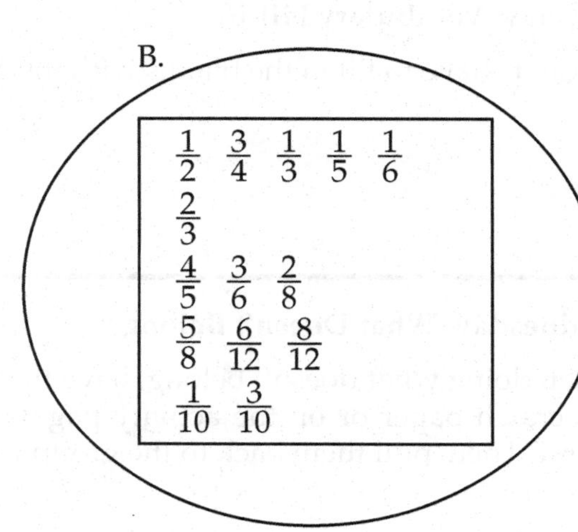

Friday: Make Your Own Problem!

Write a multi-step word problem using fractions or decimals.

Week 17 Teacher Notes

Monday: Input/Output Table

Students must fill in the empty boxes according to the rule.

Tuesday: Vocabulary Fill-in

Students have to fill in the blanks with the appropriate word.

Wednesday: What Doesn't Belong?

When doing what doesn't belong, have the students do the calculations (in their journals, on scratch paper or on the activity page). Then, have them share their thinking with a friend. Then, pull them back to the group.

Thursday: Number Strings

Students have to think about the expression and discuss which strategies they would use to solve it.

Friday: Equation Match

Students have to match the equation to the problem.

Week 17 Activities

Monday: Input/Output Table

What's the rule?

In	Out
$\frac{2}{3}$	$1\frac{1}{3}$
2	$2\frac{2}{3}$
$\frac{3}{3}$	
$\frac{10}{3}$	
	6

What's the rule?

In	Out
$\frac{1}{2}$	$\frac{1}{4}$
$\frac{1}{8}$	$\frac{1}{16}$
$\frac{1}{3}$	$\frac{1}{16}$
$\frac{1}{4}$	
	$\frac{1}{10}$

Make your own fraction pattern.

In	Out

Tuesday: Vocabulary Fill-in

Fill in the blanks with the appropriate word.

a. The answer to an addition problem is a __________

b. The answer to a subtraction problem is __________

c. The answer to a division problem is a __________

d. The answer to a multiplication problem is __________

Wednesday: What Doesn't Belong?

Students have to pick the numbers or expression that does not belong.

.25 – .20	.91 – .86
.16 – .9	.30 – .25

Thursday: Number Strings

Solve. Think money.

$2 \times .25$

$2 \times \frac{2}{4}$

$2 \times \frac{3}{4}$

$8 \times .75$

Friday: Equation Match

Sue made a blanket that was $2\frac{1}{4}$ by $3\frac{1}{2}$ feet long. How much material did she need to buy?

A. $2\frac{1}{4} + 3\frac{1}{2}$

B. $3\frac{1}{2} - 2\frac{1}{4}$

C. $2\frac{1}{4} \times 3\frac{1}{2}$

D. None of the above.

Week 18 Teacher Notes

Monday: Why Is It Not?

Students have to reason about the numbers in the equations and decide what makes the equation true.

Tuesday: Convince Me!

This routine is about getting students to defend and justify their thinking. Be sure to emphasize the language of reasoning. Students should focus on proving it with numbers, words and pictures. They should say things like:

This is true! I can prove it with …
This is the difference because …
I am going to use __________ to show my thinking.
I am going to defend my answer by __________.

Wednesday: Fraction Bingo

Students put these numbers on their bingo board. You will call them out. They will mark them on their board. Whoever gets 4 in a row first wins.

$\frac{1}{2}$ $\frac{3}{4}$ $\frac{5}{7}$ $\frac{3}{9}$ $\frac{4}{4}$ $\frac{16}{8}$ $\frac{9}{8}$ $\frac{30}{5}$
$\frac{2}{4}$ $\frac{4}{6}$ $\frac{2}{10}$ Free Space $\frac{1}{16}$ $\frac{3}{8}$ $\frac{11}{5}$. $\frac{8}{9}$

Students populate the board and then play bingo. You call out the fractions with these clues:
A fraction that is equivalent to $\frac{5}{10}$.
A fraction that is equivalent to $\frac{6}{8}$.
A Fraction that is equivalent to $\frac{1}{3}$.
A fraction that is equivalent to 1.
A fraction that is equivalent to 2.
A fraction that is a little bit more than 1 whole.
A fraction that is equivalent to 6.
A fraction that is equivalent to $\frac{7}{14}$.
A fraction that is equivalent to $\frac{2}{3}$.
A fraction that is equivalent to $\frac{1}{5}$.
A fraction close to zero.
A fraction close to half.
A fraction a little bit more than 2.
A fraction close to 1 whole.

Be sure to mix up the order of how you call them out. You can also make up your own clues for the fractions.

Thursday: Number Talk

Students have to think about the expression and discuss which strategies they would use to solve it. You want students to think about breaking the number apart.

Friday: What's the Question? (3 Read Protocol)

The focus of today is to do a 3 read problem with the students. It is important to read the problem 3 times out loud as a choral read with the students.
First read: (Stop and visualize! What do you see?) What is this story about? Who is in it? What are they doing?
Second read: What are the numbers? What do they mean?
Third read: What are some possible questions we could ask about this story?
After the students read it, write down, make up questions and then solve them. They should discuss their answers with the class.

For example: How far did he run on Monday and Tuesday?
Which day did he run the shortest distance?
How far did he run altogether?

Week 18 Activities

Monday: Why Is It Not?

a. $60 - 10 \times 2 = ?$ Why is it not 100?

b. $0.9 \times \_\_\_ = 0.54$ Why is it not 0.6?

Tuesday: Convince Me!

Convince me that a square is a rectangle but a rectangle is always not a square.

Explain why using geometry vocabulary.

Wednesday: Fraction Bingo

Put these numbers on the bingo board in different spaces. Do not put them in the order they appear. Listen as your teacher calls out equivalent fractions and cross out that space. Whoever gets 4 in a row horizontally, vertically or as a postage stamp (4 in a corner) first wins.

$\frac{1}{2}$ $\frac{3}{4}$ $\frac{5}{7}$ $\frac{3}{9}$ $\frac{4}{4}$ $\frac{16}{8}$ $\frac{9}{8}$ $\frac{30}{5}$ $\frac{2}{4}$ $\frac{4}{6}$ $\frac{2}{10}$ $\frac{3}{3}$ $\frac{1}{16}$ $\frac{3}{8}$ $\frac{11}{5}$ $\frac{8}{9}$

Thursday: Number Talk

What are some ways to think about $\frac{700}{15}$?

Friday: What's the Question? (3 Read Protocol)

Read this problem with your class 3 times. The first time talk about the story. The second time talk about the numbers. The third time, ask questions based on the story. Answer 2 of them.

Marcus ran $\frac{1}{4}$ of a mile on Monday. He ran $\frac{2}{3}$ of a mile on Tuesday. He ran $2\frac{1}{2}$ miles on Wednesday.

Week 19 Teacher Notes

Monday: Alike and Different

Students have to discuss the figures, talking about what is alike and what is different.

Tuesday: Vocabulary Bingo

Students put the words on their bingo board. Make sure they put them all over the board and not in the order listed. You will call them out using definitions or showing examples. Be sure to mix up the order. They will mark them on their board. Whoever gets 4 in a row first wins.

Wednesday: How Many More to

Students have to write how many more to each number from the start number.

Thursday: British Number Talk

Give students about 5 minutes to work on their own or with partners to come up with some problems, hopefully from each category. This is so they stretch. You don't want them to only choose the easy problem. Then, students should share what they did with class.

Friday: What's the Problem?

Students have to write a division story where the quotient is in between 75 and 100.

Week 19 Activities

Monday: Alike and Different

Discuss the figures with a math partner, talking about what is alike and what is different.

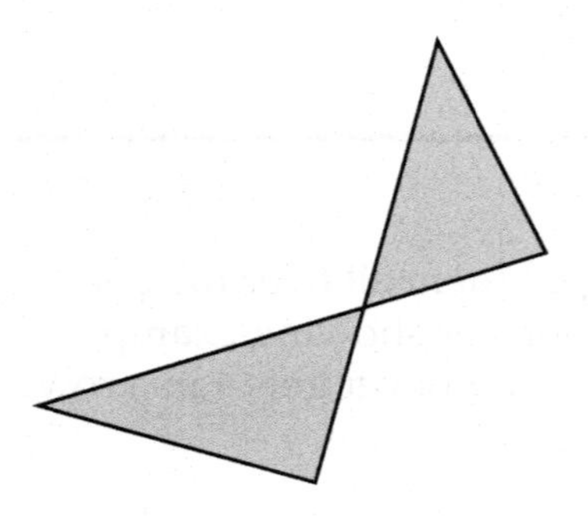

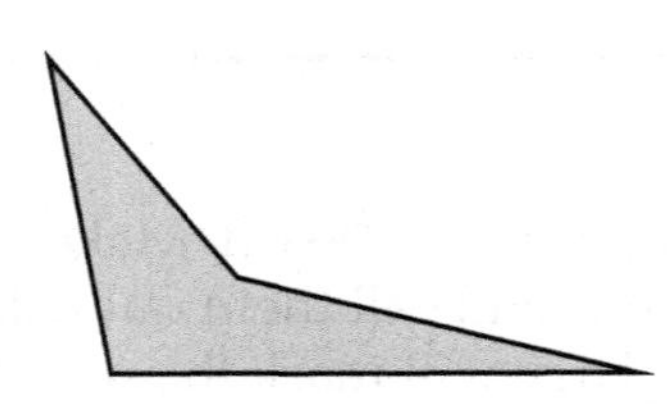

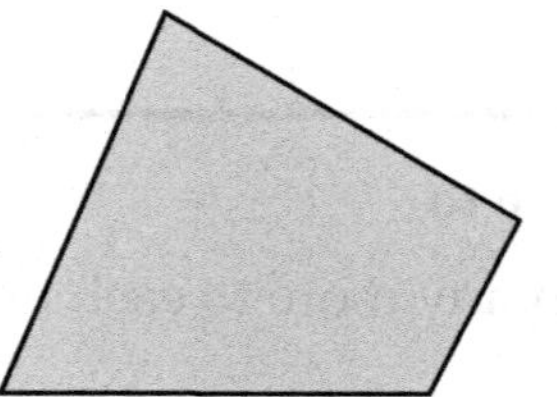

Tuesday: Vocabulary Bingo

Write a word or phrase in each space. Be sure to write them all over the board, not in the order listed. Your teacher will call out the definitions or show a picture. Whoever gets 4 in a row horizontally, vertically or as a postage stamp (4 in a corner) first wins.

Words: times, as many, comparison, exponent, multiple, algorithm, area model, open array, base of an exponent, benchmark fraction, parenthesis, equation, expression, decimal, fraction, whole number.

Wednesday: How Many More to

Think about how to get from the starter number to the designated number. Describe your thinking to your partner.

$2\frac{1}{2}$

A. Start at $\frac{4}{2}$

B. Start at $\frac{5}{2}$

C. Start at $\frac{3}{4}$

D. Start at $\frac{5}{8}$

E. Start at $\frac{0}{5}$

Thursday: British Number Talk

Pick a number from each circle. Multiply. Write the expression under the way you solved it. For example, $3 \times 3\frac{1}{4}$. I can do that in my head. 9 plus $\frac{3}{4}$ is $9\frac{3}{4}$.

I can do it in my head.	I can do it with a model.	I can do it using a written strategy or algorithm.

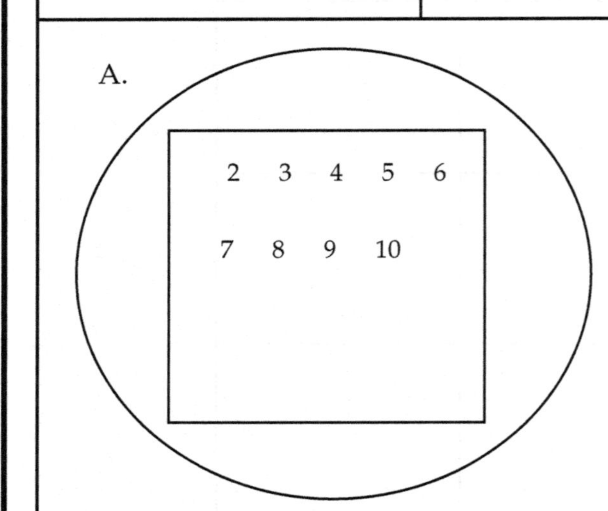

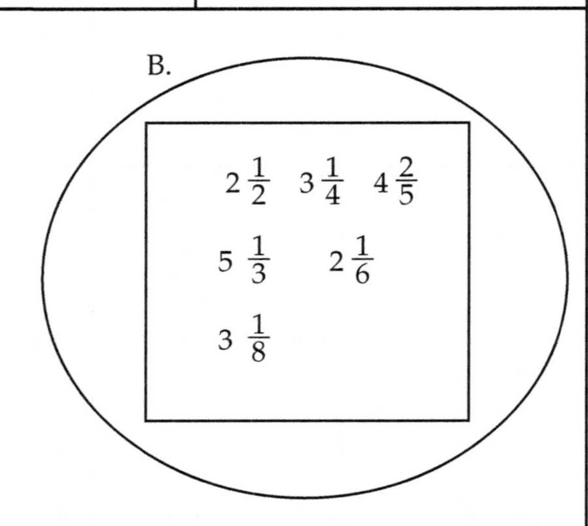

Friday: What's the Problem?

Write a division story where the quotient is in between 75 and 100.

Story:

Model:

Equation:

Week 20 Teacher Notes

Monday: True or False?

Students read the statements and decide if they are true or false. Students write that in the box. In the third column, they make their own statement.

Tuesday: Vocabulary Brainstorm

Students have to fill in information about the targeted word.

Wednesday: Patterns/Skip Counting

Create skip counting patterns. Students start where they want and create 5 terms for each of the numbers.

Thursday: Number Talk

This is a typical number talk where students are thinking about the ways in which they can solve this subtraction problem. You want students to think about partial sums, counting up, and compensation. Students are focusing on different ways. You want students to think about making friendly numbers.

$2\frac{3}{4} - \frac{4}{4}$ is taking away 1.

$3\frac{3}{4} - \frac{6}{8}$ Think about equivalent fractions.

$4\frac{2}{4} - 2\frac{3}{4}$. Think about counting up $\frac{1}{4}$ to make an easier problem: $4\frac{3}{4} - 3$.

$5\frac{2}{8} - \frac{7}{8}$ Think about counting up $\frac{1}{8}$ to make an easier problem: $5\frac{3}{8} - 1$.

Friday: What's the Story? Here's the Model

This is a fifth grade standard. The idea is that students can solve this with a visual prompt and be able to discuss its meaning.

Week 20 Activities

Monday: True or False?

	True or False?
$5 \times 7 = \frac{35}{1}$	
$5 \times 6 \times 3 = 2 \times 30$	
$\frac{2}{2} = \frac{10}{10}$	
$\frac{3}{6} > \frac{1}{2}$	
Make your own!	

Tuesday: Vocabulary Brainstorm

In each thought cloud write or draw something that has to do with equations.

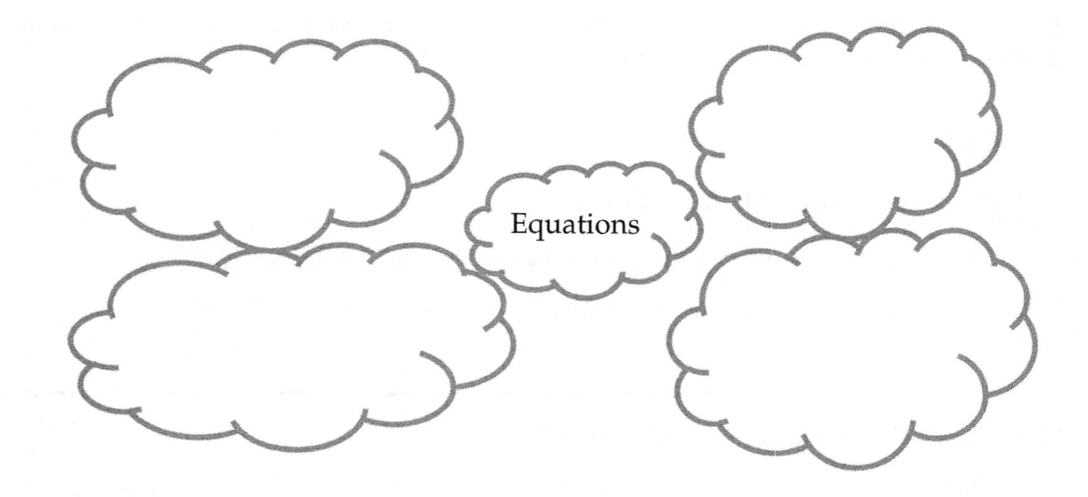

Wednesday: Patterns/Skip Counting

Start where you want and write 5 terms....

A. Count by .05.

B. Count by $\frac{4}{5}$.

C. Count by $1\frac{1}{2}$.

Thursday: Number Talk

What are some ways to think about these expressions?

1 $\frac{3}{4} - \frac{4}{4}$

2 $\frac{3}{4} - \frac{6}{8}$

3 $\frac{2}{4} - 2\frac{3}{4}$

4 $\frac{2}{8} - \frac{7}{8}$

Friday: What's the Story? Here's the Model

Write a story about fractions based on this picture.

Week 21 Teacher Notes

Monday: True or False?

Students have to think about whether or not the problems are true or false. They have to defend their thinking to their math partner and then be ready to discuss it in the whole group.

Tuesday: Vocabulary Bingo

Students have to put the words in different spaces. The teacher calls the words by stating a definition or showing a picture. When the teacher gives the definition or shows a picture, students cover the correct word. Whoever gets 4 in a row vertically, horizontally or 4 corners first wins.

Wednesday: Number Bond It!

Students have to show how to break apart a kilogram in 3 different ways. The note that 1,000 grams makes a kilogram is there so that students can think about different ways to break that apart.

Thursday: British Number Talk

Give students about 5 minutes to work on their own or with partners to come up with some problems, hopefully from each category. This is so they stretch. You don't want them to only choose the easy problem. Then, students should share what they did with class.

Friday: What's the Story? Here's the Model

Students are given the answer. They have to write a division story.

Week 21 Activities

Monday: True or False?

Think about these problems. Discuss with your partner. Be ready to share your ideas with the group.

A. $.05 < .019$

B. $.10 > .099$

C. $.20 = .200$

Tuesday: Vocabulary Bingo

Put the words on the board in any order. Your teacher will play bingo. When your teacher gives the example or shows the picture you cross it out. Whoever gets 4 in a row horizontally, vertically or as a postage stamp (4 in a corner) first wins.

Words: conversion, decimal, ones, tenths, hundredths, thousandths, decimal point, equivalent, fraction, numerator, denominator, whole number, millions, billions, ten thousands, thousands.

Write a word in each space.

Wednesday: Number Bond It!

Show how to break apart 1 kilogram in 3 different ways! 1,000 grams = 1 kilogram.

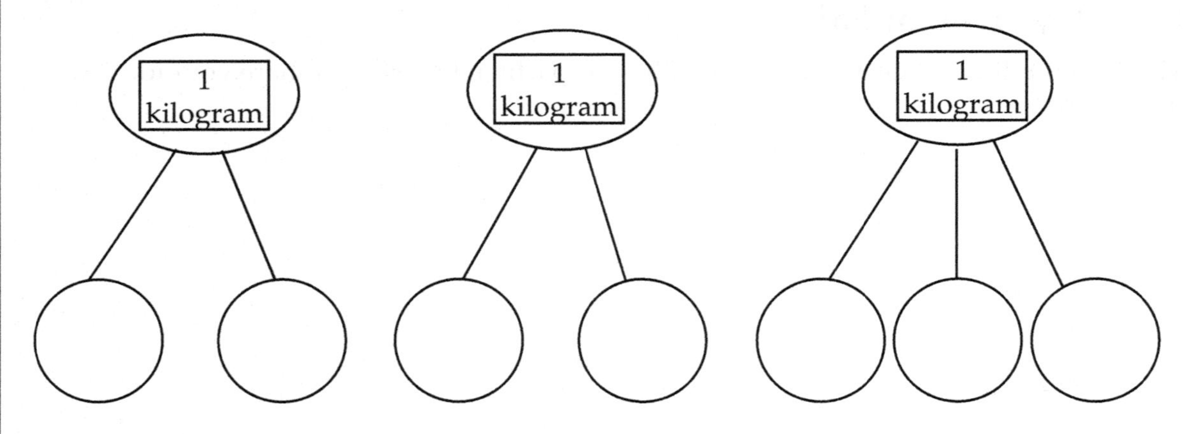

Thursday: British Number Talk

Pick a number from each circle. Add. Write the expression under the way you solved it. For example, 1.1 + .11. I can do that in my head.

I can do it in my head.	I can do it with a model.	I can do it using a written strategy or algorithm.

A.

1.1 1.2 1.3 1.4

.15 .16 .17 .18

.019 .010

B.

1.10 1.20 1.33

.14 1.54

16.1 .17 18.2

.19 .10

Friday: What's the Story? Here's the Model

The remainder is $.02. It was a division story. What was the story?

Story:

Model:

Equation:

Week 22 Teacher Notes

Monday: Missing Numbers

Students have to fill in the correct numerators so that the equation is true.

Tuesday: Vocabulary Match

Students have to match the words and the numbers.

Wednesday: Fraction of the Day

Fill in the boxes to represent the fraction.

Thursday: Number Talk

In this number talk you want the students to discuss their thinking with strategies and models. Ask students about the strategies that they might use.

Friday: Graph Talk

Students have to read the graph and then make up a story about what the data represents and tell it to their partner. They have to explain their thinking and be ready to share their thinking with the whole group.

Week 22 Activities

Monday: Missing Numbers

Talk about possible numbers that could be the numerators.

$$\frac{\square}{8} \;\_\_\_\; \frac{\square}{4} = \frac{2}{8}$$

Tuesday: Vocabulary Match

Match the words to the value of the underlined digits in the numbers.

thousandths	$\frac{4}{5}$
hundredths	5
tenths	$.93\underline{2}$
whole number	$1.0\underline{8}$
fraction	$.\underline{7}60$

Wednesday: Fraction of the Day

$2\frac{1}{2}$

Word form	Picture form
Plot it on a number line ⟵⟶	$2\frac{1}{2} <$ ______ $2\frac{1}{2} =$ ______ $2\frac{1}{2} >$ ______

Thursday: Number Talk

What are some ways to think about and show:

___ + ___ = $2\frac{1}{2}$ ___ − ___ = $2\frac{1}{2}$ ___ × ___ = $2\frac{1}{2}$

Friday: Graph Talk

Here's the line plot.	Read the graph. Write a story to match the graph. Ask and answer 3 different questions about the data.

Here's the line plot.

$\frac{1}{8}$	$\frac{2}{8}$	$\frac{1}{2}$	$\frac{3}{4}$	$1\frac{1}{2}$	$1\frac{3}{4}$	$2\frac{1}{8}$
						x
				x		x
		x		x		x
		x		x	x	x
x		x		x	x	x
x		x		x	x	x

Read the graph. Write a story to match the graph. Ask and answer 3 different questions about the data.

Week 23 Teacher Notes

Monday: Magic Square

Students have to fill in the squares with numbers so that all directions end in a sum of 9.

Tuesday: Frayer Model

Students have to fill in the different boxes to represent the word.

Wednesday: Number Line It!

Students have to order the numbers from least to greatest on the number lines and explain their thinking.

Thursday: Number Talk

Students discuss ways to solve the expression.

Friday: Model That!

Students have to model the problem with a tape diagram.

Week 23 Activities

Monday: Magic Square

Fill in the squares with numbers. The sum must total 9 in all directions.

Sum is 9

3.3		3.1
	3.0	

Tuesday: Frayer Model

Fill in the boxes based on the word.

Powers of ten

Definition	Examples
Give a picture example	Non-examples

Wednesday: Number Line It!

Order these numbers from least to greatest on the number lines. Be ready to explain your thinking.

A. $\frac{4}{5}$ $\frac{7}{9}$ $2\frac{1}{2}$ $\frac{7}{4}$ $\frac{0}{9}$

B. .34 1.02 .59 .17 .01

C. Draw a number line and plot $\frac{5}{3}$ and 2 fractions that are less than $\frac{5}{3}$ and 2 fractions that are more than $\frac{5}{3}$.

Thursday: Number Talk

What are some ways to solve $567 \div 14$?

Friday: Model That!

In the tank there were 15 animals. $\frac{1}{5}$ of them were turtles, 2 times that amount were eels and the rest were small orange fish. How many animals were turtles? How many animals were eels? How many were orange fish?

Model it with a tape diagram.

Week 24 Teacher Notes

Monday: Input/Output Table

Students have to create an input output table where the rule is to multiply by $\frac{2}{3}$.

Tuesday: 1-Minute Essay

Students have 1 minute to write everything they can about fractions using numbers, words and pictures.

Wednesday: Rounding

Students have to name 3 numbers that round to the designated number.

Thursday: British Number Talk

Give students about 5 minutes to work on their own or with partners to come up with some problems, hopefully from each category. This is so they stretch. You don't want them to only choose the easy problem. Then, students should share what they did with class.

Friday: Model That!

Students have to read, model and solve the word problem.

Week 24 Activities

Monday: Input/Output Table

Create an input/output table where the rule is to multiply by $\frac{2}{3}$.

In	Out

Tuesday: 1-Minute Essay

You have 1 minute to write everything you know about fifth grade math.

Go!

Wednesday: Rounding

Name 3 decimals that round to .8.

Thursday: British Number Talk

Pick a number from each circle. Divide them. Decide how you will solve it and write that expression under the title.

I can do it in my head.	I can do it with a model.	I can do it using a written strategy or algorithm.

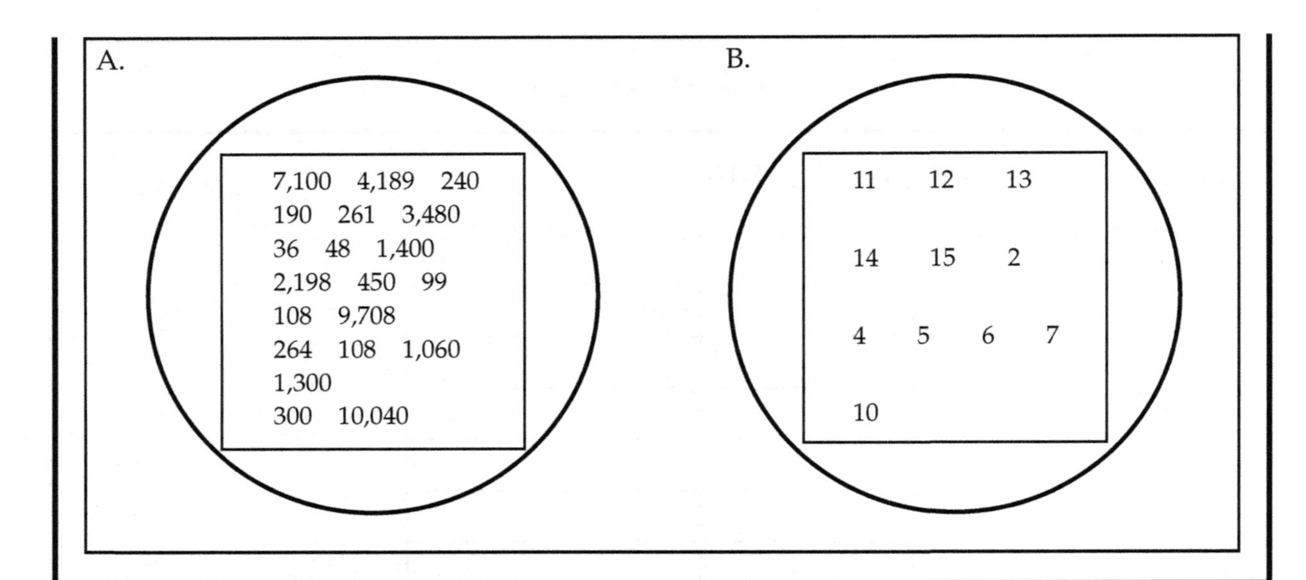

Friday: Model That!

At the butterfly museum $\frac{1}{3}$ of the insects were butterflies, $\frac{3}{6}$ of the insects were ladybugs and the rest were grasshoppers. What fraction of the insects are grasshoppers? There were 12 insects altogether. How many of each animal were there?

Model the problem.

Week 25 Teacher Notes

Monday: What Doesn't Belong?

When doing this routine, have the students do the calculations (in their journals, on scratch paper or on the activity page). Then, have them share their thinking with a friend. Then, pull them back to the group.

Tuesday: Vocabulary Tic Tac Toe

These are quick partner energizers. Read all the words together. Then go! They have 7 minutes to play the game. They do rock, paper, scissors to start. They take turns choosing a word and explaining it to their partner. Then, they have to do a sketch or something to show they understand the word. Everybody should play the first game, if they have time, they can play the next one.

Note: It is important to call everyone back together at the end and talk about the vocabulary. Briefly go over the vocabulary.

Wednesday: Decimal of the Day

Students have to fill in the boxes to represent the fraction.

Thursday: Number Talk

In this number talk you want the students to discuss their thinking with strategies and models. Ask students about the strategies that they might use.

Friday: Make Your Own Problem!

Students have to read, model and solve the word problem.

Week 25 Activities

Monday: What Doesn't Belong?

Look at the different sets. Pick the numbers that do not belong in each set. Explain why the number doesn't belong to your math partner.

A.

$.21 - .14$	$.45 - .38$
$.16 - .09$	$.30 - .25$

B.

$(3 \times 6) + (3 \times 1)$	$(3 \times 3) + (3 \times 4)$
$(3 \times 5) + (2 \times 4)$	3×7

Tuesday: Vocabulary Tic Tac Toe

Read all the words together. Then go! You have 7 minutes to play the game. Do rock, paper, scissors to start. Then, take turns choosing a word and explaining it to your partner. Then, do a sketch or something to show that you understand the word. Everybody should play the first game, if you have time, they can play the next one.

>	<	=
tenth	hundredth	thousandth
place value	decimal	digit

equation	exponent	powers of ten
expression	inequality	comparison
represent	operation	equivalent

Wednesday: Decimal of the Day

Fill in the boxes correctly based on the number.

.078

Word form	___ + ____ = .078 _____ − _____ = .078 $.078 = \_\_\_ \times \frac{1}{10} + \_\_\_ \times \frac{1}{100} + \_\_\_ \times \frac{1}{1000}$
.078 < ____ .078 > ___	Round it to the nearest hundredth _____ Round it to the nearest tenth _______

Thursday: Number Talk

How can we solve $3 \times .75$?

Friday: Make Your Own Problem!

Write a story about this expression.

$112 \div 12$

Model your thinking.

Week 26 Teacher Notes

Monday: Reasoning Matrices

Students have to reason about the problem and be ready to discuss it.

Tuesday: Vocabulary Fill-in

Students have to read the sentences and then fill in the blanks with the appropriate words.

Wednesday: Greater Than, Less Than, in Between

Students have to read the descriptions and fill in appropriate numbers.

Thursday: Find and Fix the Error

Students have to look at and think about the problem. They decide what is wrong and fix it. Then they discuss the work with a math partner.

Friday: What's the Story? Here's the Model

Students have to write a story to match the model.

Week 26 Activities

Monday: Reasoning Matrices

Which of these will equal a number less than 1,000? Which of these will equal a number greater than 1,000? Which of these will equal a number close to 1,000?

a. $384 \times 1{,}000$

b. $0.88 \times 1{,}000$

c. $\frac{1}{3} \times 1{,}000$

d. $\frac{13}{4} \times 1{,}000$

Explain your thinking to a neighbor. Be ready to discuss it with the whole class.

Tuesday: Vocabulary Fill-in

Students have to read the sentences and then fill in the blanks with the appropriate words.

Miles and kilometers measure ____________.

We use pints, quarts, and liters to measure ________.

Volume measures ________________.

Mass and weight are measured with _____ and _______.

Wednesday: Greater Than, Less Than, in Between

Students have to read the descriptions below and fill in appropriate numbers.

.03 .45 2.79

Name a decimal that is greater than to .03	Name a decimal greater than .45	Name a decimal less than 2.79
Name a decimal that is smaller than .03	Name a decimal in-between .03 and .45	Name a decimal in-between .45 and 2.79

Thursday: Find and Fix the Error

Mike solved the problem this way:

$\frac{1}{2} \times \frac{1}{2} = \frac{2}{4}$

Why is it wrong?

What did he do wrong?

How can we fix it?

Friday: What's the Story? Here's the Model

Look at the tape diagram below. Write a story that matches the tape diagram.

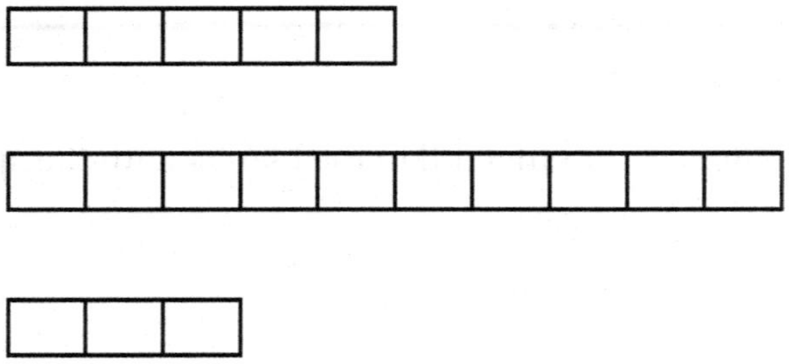

Story:

Equation:

Week 27 Teacher Notes

Monday: 3 Truths and a Fib

Read the boxes. 3 are true and one is false. Which one is false?

Tuesday: Vocabulary Talk

Students have to talk about the relationships between the words.

Wednesday: Input/Output Table

Students have to fill in the in/out table following the given rule.

Thursday: Number Talk

This is a typical number talk where students are thinking about the ways in which they can solve this division problem.

Friday: Equation Match

Students have to match the problem with the equation.

Week 27 Activities

Monday: 3 Truths and a Fib

Read the boxes. 3 are true and one is false. Which one is false?

A.

5 is composite	8 is not prime
20 is composite	12 is not prime

B.

.06 > .006	.90 < .105
.405 is in between .407 and .399	.80 = .8000

Tuesday: Vocabulary Talk

Discuss these word pairs with your partner.

equation/inequality

numerator/denominator

multiple/factor

Wednesday: Input/Output Table

Create a table using this rule: divide by .02.

In	Out

Thursday: Number Talk

What are some ways to solve $6 \div \frac{1}{2}$?

Friday: Equation Match

Match the equation to the correct story. Explain your thinking. Solve.

$$\frac{2}{5} - a = \frac{1}{10}$$

Problem A	Problem B
Aunty Mary had $\frac{2}{5}$ cup of butter. She used $\frac{1}{10}$. How much does she have left?	Aunty Mary had $\frac{2}{5}$ cup of butter. She used some. Now, she has $\frac{1}{10}$ of a cup of butter left. How much did she use?

Week 28 Teacher Notes

Monday: Why Is It Not?

Students look at each problem and think about why the given answer is not correct. They have to discuss it and then explain their thinking to the class.

Tuesday: It Is/It Isn't

In this routine you want students to be focusing on the vocabulary. Encourage students to use the word bank. This is a scaffold only though.

Wednesday: How Many More to

The students have to reason about how to get from the starter number to the designated number.

Thursday: British Number Talk

Give students about 5 minutes to work on their own or with partners to come up with some problems, hopefully from each category. This is so they stretch. You don't want them to only choose the easy problem. Then, students should share what they did with class.

Friday: What's the Question? (3 Read Protocol)

The focus of today is to do a 3 read problem with the students. It is important to read the problem 3 times out loud as a choral read with the students.

First read: (Stop and visualize! What do you see?) What is this story about? Who is in it? What are they doing?

Second read: What are the numbers? What do they mean?

Third read: What are some possible questions we could ask about this story? After the students read it, write down, make up questions and then solve them. They should discuss their answers with the class. For example: How many pies were lemon? How many pies were strawberry?

Note: Focus on the vocabulary.

Focus on different types of comparative language so students get comfortable with words and phrases like: How many more? How much less? How many more to get the same amount as? How many fewer?

Week 28 Activities

Monday: Why Is It Not?

Think about it and share it with a partner. Share with the class.

$\frac{2}{12} + \_\_\_\_ = \frac{3}{6}$

Why is it not $\frac{5}{18}$?

Tuesday: It Is/It Isn't

Describe a right angle using the frame It Is/It Isn't.

A right angle

It Is	It Isn't

Vocabulary bank: more than, less than, greater than, degrees, obtuse, straight, right angle.

Wednesday: How Many More to

Think about how to get from the starter number to the designated number. Describe your thinking to your partner.

A. Start at .39 and get to 1.

B. Start at 1.13 and get to 2.

C. Start at 0.6 and get to 4.

D. Make up your own start at/get to problem.

Thursday: British Number Talk

Pick a number from each circle. Multiply. Write the expression under the way you solved it. For example, $\frac{1}{2} \times \frac{1}{2}$. I can do it in my head.

I can do it in my head.	I can do it with a model.	I can do it using a written strategy or algorithm.

A.

$\frac{1}{2}$ $\frac{2}{3}$ $\frac{3}{4}$

$\frac{4}{7}$ $\frac{5}{8}$ $\frac{6}{9}$

$\frac{5}{6}$ $\frac{2}{5}$

B.

$\frac{1}{2}$ $\frac{2}{3}$ $\frac{3}{4}$

$\frac{4}{5}$ $\frac{5}{10}$

$\frac{6}{8}$ $\frac{2}{6}$ $\frac{4}{12}$

$\frac{0}{6}$

Friday: What's the Question? (3 Read Protocol)

Read the problem 3 times. The first time talk about the situation. The second time talk about the numbers. The third time, ask questions based on the situation. Answer 2 of them.

Grandma Betsy made 10 pies. $\frac{1}{5}$ of them were lemon, $\frac{2}{10}$ of them were strawberry. There were also some peach ones.

Week 29 Teacher Notes

Monday: Guess My Number

Students have to read the clues and guess the number.

Tuesday: Vocabulary Bingo

Students have to put these words on the bingo board in different spaces. They should not put them in the order they appear. They listen to the teacher call out the definitions or show a picture. When they hear it they cross out the word. Whoever gets 4 in a row first wins.

Wednesday: Find and Fix the Error

Students have to read and discuss the work. They have to figure out what is incorrect and discuss how to fix it.

Thursday: Number Talk

This is a typical number talk where students are thinking about the ways in which they can solve this problem.

Friday: What's the Problem?

Students have to write a word problem where the quotient is about 8.

Week 29 Activities

Monday: Guess My Number

Read the clues. Guess the number.

A.	B.
I am thinking of a decimal that is greater than $2 \times .05$. I am thinking of a number that is less than $3 \times .07$. It is a multiple of .09. What is my decimal?	I am thinking of a fraction that is greater than $\frac{1}{2}$. I am thinking of a fraction that is less than $\frac{2}{2} \times \frac{3}{2}$. It is equivalent to $\frac{9}{12}$. What is my fraction?

Tuesday: Vocabulary Bingo

Put these words on the bingo board in different spaces. Do not put them in the order they appear. Listen as your teacher calls out the definitions or shows a picture. When you hear it cross out the word. Whoever gets 4 in a row horizontally, vertically or as a postage stamp (4 in a corner) first wins.

Words: endpoint, product, angle, acute, obtuse, straight, right angle, parallel, perpendicular, ray, intersecting, degree, line, line segment, parallelogram, figure.

Write a word in each space.

Wednesday: Find and Fix the Error

Joe solved the problem below. He made an error.

Find his error. Discuss it with your partner.

Fix it.

$$\begin{array}{r} .778 \\ +\ .54 \\ \hline .12118 \end{array}$$

Thursday: Number Talk

Write expressions that equal $5\frac{1}{2}$ by using addition, subtraction, multiplication and division.

Friday: What's the Problem?

Write a word problem where the quotient is about 8.

Word problem:

Model:

Equation:

Week 30 Teacher Notes

Monday: Open Array Puzzle

Students have to solve the problem with an open array. They fill in the puzzle to solve the problem.

Tuesday: Vocabulary Talk

Students have to talk about the relationship between the designated words.

Wednesday: Money Mix Up

Students have to show and discuss how they make the designated amount.

Thursday: British Number Talk

Give students about 5 minutes to work on their own or with partners to come up with some problems, hopefully from each category. This is so they stretch. You don't want them to only choose the easy problem. Then, students should share what they did with class.

Friday: What's the Problem?

Students have to write a word problem about volume. The answer is 60 in^3.

Week 30 Activities

Monday: Open Array Puzzle

Fill in the open array puzzle.

$\frac{787}{8}$

__________ + __________

?	?

Answer is _______ remainder __

Tuesday: Vocabulary Talk

Talk about the relationship between mm, cm, m, km

Talk about the relationship between mg, g, kg

Talk about the relationship between cup, pint, quart, half gallon, gallon

Wednesday: Money Mix Up

Ted had $5. He spent half on lollipops, $\frac{1}{5}$ on gum and split the rest with his little sister. Write how much money he spent on lollipops, gum, his sister and how much he kept.

Thursday: British Number Talk

Pick a number from each circle. Divide. Write the expression under the way you solved it. For example, .416 ÷ 7. I can do with a written strategy or algorithm.

I can do it in my head.	I can do it with a model.	I can do it using a written strategy or algorithm.

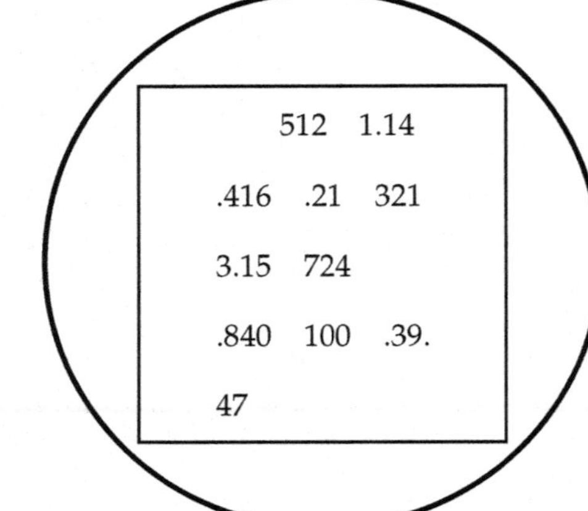

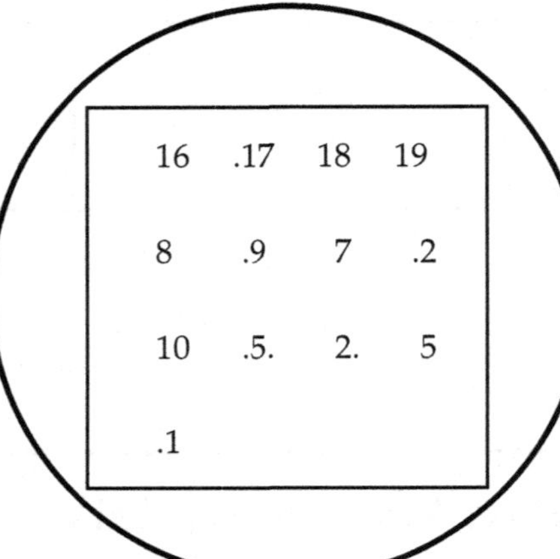

Friday: What's the Problem?

Write a word problem about volume where this is the answer.
The answer is 60 in^3.

Word problem:

Model:

Equation:

Week 31 Teacher Notes

Monday: What Doesn't Belong?

Students have to look at the set and decide which one doesn't belong. They have to explain their thinking and defend their answer to a partner.

Tuesday: Vocabulary Brainstorm

Students fill in the clouds with numbers, words and pictures to discuss the target word.

Wednesday: 3 Truths and a Fib

Students have to read the statements and decide which 3 are true and which one is not. They then discuss with their neighbor which one is false. They must explain their thinking to their neighbor and then to the class.

Thursday: Number Talk

This is a typical number talk where students are thinking about the ways in which they can solve these division problems.

Friday: What's the Story? Here's the Model

Students have to write a story about the given expression.

Week 31 Activities

Monday: What Doesn't Belong?

Look at the sets below. Which numbers do not belong in each set?

A.

$7.45 = 7\frac{45}{10}$	$3.21 = 3\frac{21}{100}$
$7.5 = 7\frac{5}{10}$	$4.3 = 4\frac{3}{10}$

B.

$15 \times 2 \times 10$	$2 \times 30 \times 5$
$30 \times 10 \times 0$	$10 \times 1 \times 30$

Tuesday: Vocabulary Brainstorm

In each thought cloud write or draw something that has to do with the given term.

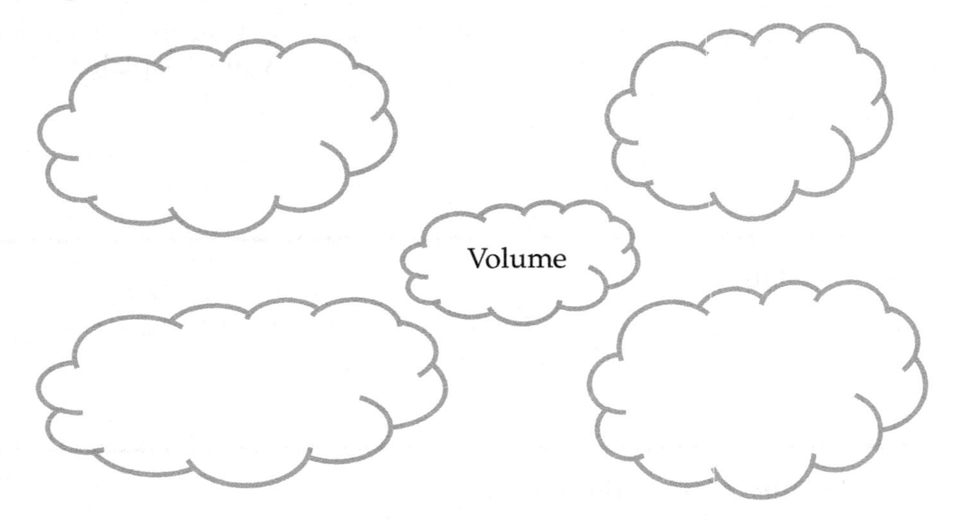

Wednesday: 3 Truths and a Fib

Three of the problems are correct and 1 is false in each set. Find the false one.

A.	B.
1. $\frac{.25}{.05} = 5$	1. $\frac{1}{3} = \frac{2}{6}$
2. $\frac{.75}{.25} = 3$	2. $\frac{4}{5} = \frac{8}{20}$
3. $\frac{1}{.25} = 4$	3. $\frac{2}{5} = \frac{4}{10}$
4. $\frac{1}{.20} = 4$	4. $\frac{1}{4} = \frac{2}{8}$

Thursday: Number Talk

Discuss these problems.

$2.15 \div 10$

$2.15 \div 100$

$2.15 \div 1000$

Friday: What's the Story? Here's the Model

Look at the expression below. Write a story to match the expression.

$\frac{1}{2} \div 4$	

Story:

Model:

Week 32 Teacher Notes

Monday: Open Array Puzzle

Students have to solve the problem using an open array. They fill in the correct numbers in the open array.

Tuesday: What Doesn't Belong?

Students have to decide which word doesn't belong in each set.

Wednesday: 10 Times as Much

Student have to follow the directions and write the correct numbers.

Thursday: British Number Talk

Give students about 5 minutes to work on their own or with partners to come up with some problems, hopefully from each category. This is so they stretch. You don't want them to only choose the easy problem. Then, students should share what they did with class.

Friday: What's the Story? Here's the Model

Students have to tell the story about the tape diagram.

Week 32 Activities

Monday: Open Array Puzzle

Fill in the boxes and write the partial products.

$12 \times 18 =$

	10	+ 8
10		
+ 2		

Tuesday: What Doesn't Belong?

Which word doesn't belong in each set?

A.

acute	obtuse
straight	prime

B.

degree	digit
protractor	ray

Wednesday: 10 Times as Much

Write a number where the digit in the thousands place is 10 times the digit in the place next to it.

Explain to your neighbor why this is true. Discuss with the class.

Thursday: British Number Talk

Pick a number from each circle. Multiply. Write the expression under the way you solved it. For example, 24.6 x 10. I can do that in my head.

I can do it in my head.	I can do it with a model.	I can do it using a written strategy or algorithm.

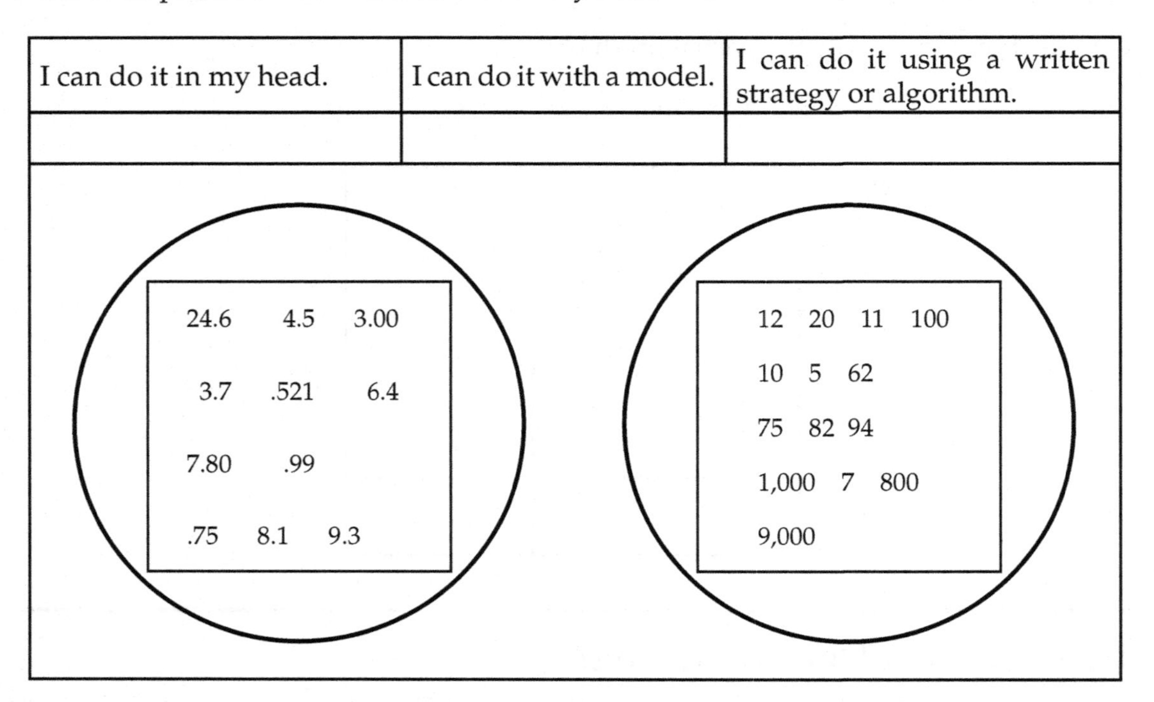

Friday: What's the Story? Here's the Model

Look at the bar diagram below. Write a story that matches the bar diagram.

50			
12.50	12.50	12.50	12.50

Story:

Equation:

Week 33 Teacher Notes

Monday: Reasoning Matrices

Students have to reason about the problem and be ready to discuss it.

Tuesday: Vocabulary Talk

Students discuss the different vocabulary words.

Wednesday: Decimal Comparison

Students have to compare and model decimals.

Thursday: Number Talk Puzzle

Students have to figure out which numbers are missing to complete the multiplication puzzle.

Friday: Graph Talk

Students have to read the graph and then tell a story and ask and answer questions about it.

Week 33 Activities

Monday: Reasoning Matrices

A. What will happen to a whole number when it is multiplied by a fraction less than 1? Give an example.
B. What will happen to a whole number when it is multiplied by a fraction greater than 1? Give an example.
C. What will happen to a whole number when it is multiplied by 1? Give an example.

Explain your thinking to a neighbor. Be ready to discuss it with the whole class.

Tuesday: Vocabulary Talk

Name 2 things you can measure the volume of.

Name 2 things you measure the area of.

Name 2 things you measure the perimeter of.

Name 2 things that you can measure in cubic feet.

Wednesday: Decimal Comparison

Write a story comparing 3 decimals. Use the grids below to model the story.

Draw a number line and order them from least to greatest. Explain your reasoning.

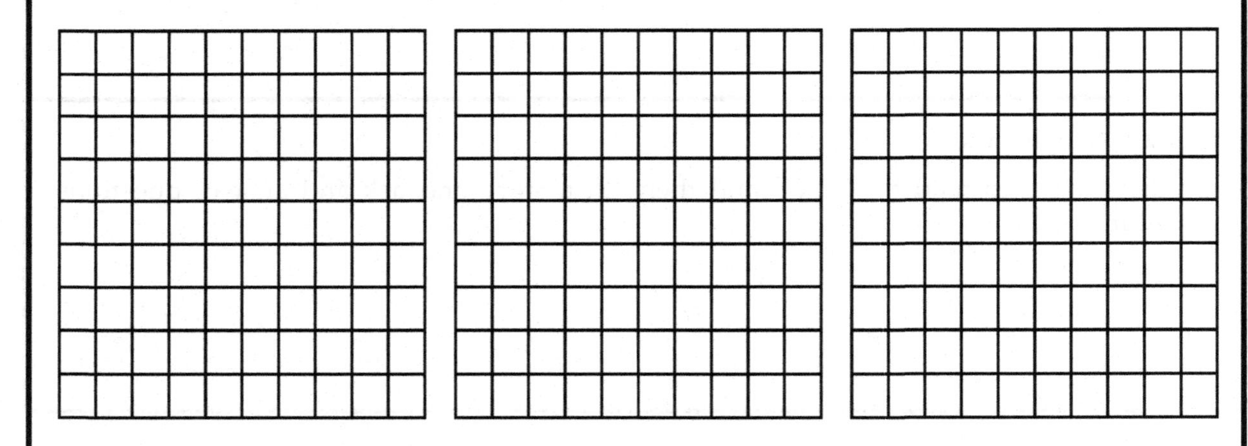

Story:

Statements:

____ > _____

______< _____

Thursday: Number Talk Puzzle

Fill in the missing numbers so that they make sense. Solve the problem.

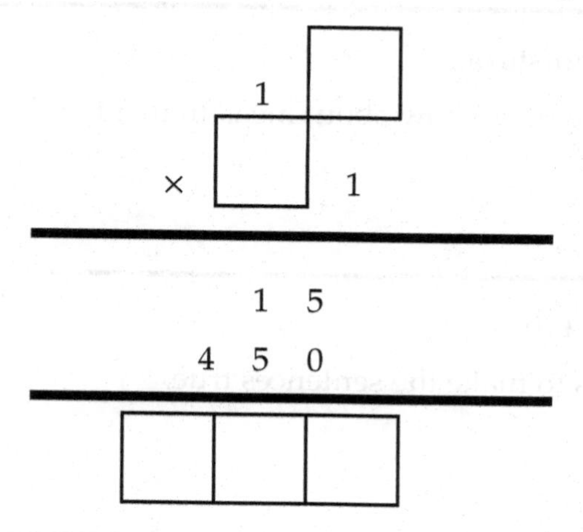

Explain your thinking to a partner.

Be able to defend your thinking in the whole group discussion.

Friday: Graph Talk

Here's the line plot.

					x
			x		x
		x	x		x
		x	x	x	x
x		x	x	x	x
x	x	x	x	x	x
$\frac{1}{2}$	1	$1\frac{1}{4}$	$1\frac{1}{2}$	$1\frac{3}{4}$	$2\frac{1}{2}$

Read the graph. Write a story to match the graph. Ask and answer 3 different questions about the data.

Week 34 Teacher Notes

Monday: Coordinate Grid

Students plot a coordinate on the grid.

Tuesday: Vocabulary Brainstorm

Students write everything they know about measurement in the boxes.

Wednesday: Vocabulary Fill-in

Students fill in the spaces to make the sentences true.

Thursday: Find and Fix the Error

Students have to find and fix the error. They have to discuss their thinking with a math partner.

Friday: Model That!

Students have to draw a model to match the problem.

Week 34 Activities

Monday: Coordinate Grid

Explain your thinking to a neighbor. Be ready to discuss it with the whole class.

Label the x and y axes on the coordinate grid. Draw a point and write down the coordinates.

Tuesday: Vocabulary Brainstorm

Write down everything that you know about measurement using numbers, words and pictures. Discuss with your math partner.

Mass	Liquid Volume	Weight	Length	Volume

Wednesday: Vocabulary Fill-in

Make true statements:

There are _____ mm in ____ meter.

There are _____ cl in _____liter.

There are ______g in ______kilogram.

Thursday: Find and Fix the Error

Kevin solved the problem this way:

$$6 \div \frac{1}{2} = \frac{6}{2} = 3$$

Why is it wrong?

What did he do wrong?

How can we fix it?

Friday: Model That!

There was a rectangle. The area was more than double the perimeter. What could it have looked like?	Model it.

Week 35 Teacher Notes

Monday: Find and Fix the Error

Students have to find and fix the error. They have to discuss it with a math partner.

Tuesday: Vocabulary Tic Tac Toe

These are quick partner energizers. Read all the words together. Then go! They have 7 minutes to play the game. They do rock, paper, scissors to start. They take turns choosing a word and explaining it to their partner. Then, they have to do a sketch or something to show they understand the word. Everybody should play the first game, if they have time, they can play the next one.

Note: It is important to call everyone back together at the end and talk about the vocabulary. Briefly go over the vocabulary.

Wednesday: Fraction of the Day

Students have to talk with a partner and then share with the class their answers.

Thursday: British Number Talk

Give students about 5 minutes to work on their own or with partners to come up with some problems, hopefully from each category. This is so they stretch. You don't want them to only choose the easy problem. Then, students should share what they did with class.

Friday: Perimeter Problem

Students have to solve this perimeter problem.

Week 35 Activities

Monday: Find and Fix the Error

Look at how Jay solved the problem.

What did he do wrong?

How can he fix it?

$$
\begin{array}{r}
38 \\
\times\ 12 \\
\hline
616 \\
+\ 380 \\
\hline
996
\end{array}
$$

Tuesday: Vocabulary Tic Tac Toe

Play rock, paper, scissors to see who goes first. Then take turns, picking a square, saying what it means, drawing or writing something about the word on the side and then marking the word with an x or an o. Whoever gets 3 in a row first wins.

acute	parallel	90 degrees
straight angle	perpendic-ular	360 degrees
obtuse angle	intersecting	180 degrees

multiple	composite	divisor
factor	factor pair	quotient
prime	dividend	expression

Wednesday: Fraction of the Day

A. Add 2 fractions that almost make 1.
B. Subtract 2 fractions that almost make $1\frac{1}{2}$.
C. Multiply 2 fractions that make $\frac{3}{4}$.
D. Make your own fraction problem and solve.

Thursday: British Number Talk

Pick a number from each circle. Subtract. Write the expression under the way you solved it. For example, 3.00 – .89. I can do it using a written strategy or algorithm.

I can do it in my head.	I can do it with a model.	I can do it using a written strategy or algorithm.

A.

.1 .25 .39
.48 .52 .63
.74 .85 .90
1.10 2.40 3.00

B.

1.99 .32 .55 .61
.72
.43 .24 .75
.98 .89 2.07

Friday: Perimeter Problem

There was a fence around a yard that had an area of 2,400 square feet. The width is 60 feet. What is the perimeter?

Week 36 Teacher Notes

Monday: What's the Problem?

Students have to think about a problem where the quotient is around .7.

Tuesday: What Doesn't Belong?

Students have to decide which words do not go with the set.

Wednesday: Guess My Number

Students have to read the riddle and use the clues to determine what the number is. They then have to discuss the answer with a classmate.

Thursday: British Number Talk

Give students about 5 minutes to work on their own or with partners to come up with some problems, hopefully from each category. This is so they stretch. You don't want them to only choose the easy problem. Then, students should share what they did with class.

Friday: What's the Story? Here's the Model

Students have to match the model with the correct story.

Week 36 Activities

Monday: What's the Problem?

Write a problem that has a quotient of .7. Try to think of 3 examples.

Tuesday: What Doesn't Belong?

Choose the word that does not belong.

A.

multiple	factor
product	difference

B.

pint	quart
gram	liter

Wednesday: Guess My Number

Follow the clues to figure out what the mystery number is.

A.	B.
I am a decimal that is less than 1. I am greater than $2 \times .37$. I am a 2 digit number. I am less than $2 \times .39$. Both my digits are the same number. Who am I?	I am a decimal that is less than $4 \times .10$. I am greater than $3 \times .11$. I am a multiple of .12. Who am I?

Thursday: British Number Talk

Pick a number from each circle. Make a multiplication problem. Write the problem under the way you solved it.

I can do it in my head.	I can do it with a model.	I can do it using a written strategy or algorithm.

A.

Make your own 2 digit numbers.

B.

Make your own 2 digit numbers.

Friday: What's the Story? Here's the Model

Which story could match this model?

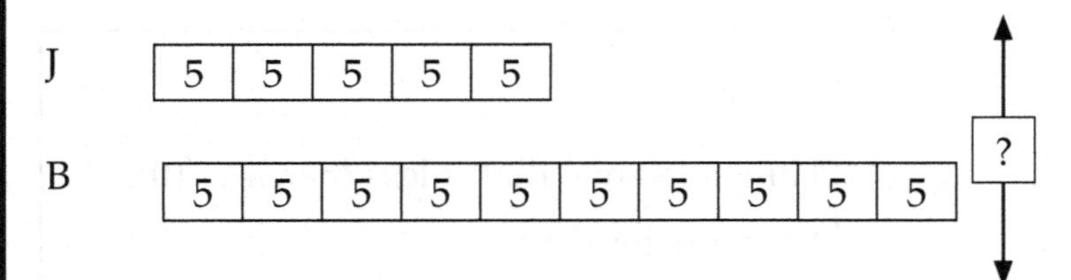

1. John had 25 marbles. His brother had twice as many as he did. How many do they have altogether?
2. John had 25 marbles. His brother had 3 times as many as he did. How many more does his brother have than he does?
3. John has 25 marbles. He has twice as many as his brother. How many do they have altogether?

Week 37 Teacher Notes

Monday: Convince Me!

This routine is about getting students to defend and justify their thinking. Be sure to emphasize the language of reasoning. Students should focus on proving it with numbers, words and pictures. They should say things like:

This is true! I can prove it with …
This is the difference because …
I am going to use __________ to show my thinking.
I am going to defend my answer by __________.

Tuesday: Vocabulary Bingo

Students put the words in the box in a different order than they appear. Play bingo. When the teacher calls a word, cover it. Whoever gets 4 in a row vertically, horizontally or 4 corners wins.

Wednesday: What Doesn't Belong?

Students have to choose the fractions that are not part of the sets.

Thursday: British Number Talk

Give students about 5 minutes to work on their own or with partners to come up with some problems, hopefully from each category. This is so they stretch. You don't want them to only choose the easy problem. Then, students should share what they did with class.

Friday: Make Your Own Problem!

In this routine students have to find the difference between 2 decimals and write them. They have to write a story to go with one of the examples.

Week 37 Activities

Monday: Convince Me!

A. $\frac{3}{4}$ is greater than $\frac{1}{8}$

B. $\frac{2}{3}$ is greater than $\frac{3}{6}$

C. $\frac{7}{6}$ is less than $\frac{4}{3}$

Tuesday: Vocabulary Bingo

Put the words in the box in a different order than they appear. Play bingo. When the teacher calls a word, cover it. Whoever gets 4 in a row vertically, horizontally or 4 corners wins.

Words: line plot, oz, lbs, quarts, pints, l, ml, cm, mm, kg, g, mg, gallons.

Wednesday: What Doesn't Belong?

Look at the fractions in each set. Decide which one does not belong. Discuss with your math partner. Be ready to defend your thinking to the group.

A.

$\frac{1}{4}$	$\frac{2}{8}$
$\frac{4}{16}$	$\frac{3}{6}$

B.

$\frac{5}{4}$	$\frac{2}{4}$
$\frac{1}{8}$	$\frac{9}{12}$

Thursday: British Number Talk

Pick a number from each circle. Subtract. Write the expression under the way you solved it.

I can do it in my head.	I can do it with a model.	I can do it using a written strategy or algorithm.

A.

Write your own fractions.

B.

Write your own fractions.

Friday: Make Your Own Problem!

A. A difference of 0.10.
B. A difference of 0.50.
C. A difference of 1.07.
D. Write a word problem about one of the above problems.

Week 38 Teacher Notes

Monday: Input/Output Table

Students have to look at the pattern and figure out what comes next. They should be ready to discuss their thinking with a math partner.

Tuesday: Vocabulary Tic Tac Toe

These are quick partner energizers. Read all the words together. Then go! They have 7 minutes to play the game. They do rock, paper, scissors to start. They take turns choosing a word and explaining it to their partner. Then, they have to do a sketch or something to show they understand the word. Everybody should play the first game, if they have time, they can play the next one.

Note: It is important to call everyone back together at the end and talk about the vocabulary. Briefly go over the vocabulary, this is all vocabulary that students should know.

Wednesday: Place Value Puzzle

Students have to reason about the numbers and make a problem based on the directions.

Thursday: British Number Talk

Give students about 5 minutes to work on their own or with partners to come up with some problems, hopefully from each category. This is so they stretch. You don't want them to only choose the easy problem. Then, students should share what they did with class.

Friday: Word Problem Fill-in

Students fill in the word problem with their own numbers and model and solve the problem.

Week 38 Activities

Monday: Input/Output Table

Make a pattern that adds .09 to the input number.

Input	Output

Tuesday: Vocabulary Tic Tac Toe

Play rock, paper, scissors to see who goes first. Then take turns, picking a square, saying what it means, drawing or writing something about the word on the side and then marking the word with an x or an o. Whoever gets 3 in a row first wins.

A.

mg	kg	gram
oz	km	quarts
miles	lbs	gallons

B.

oz	cl	cm
ml	cm	m
cm	l	kg

Wednesday: Place Value Puzzle

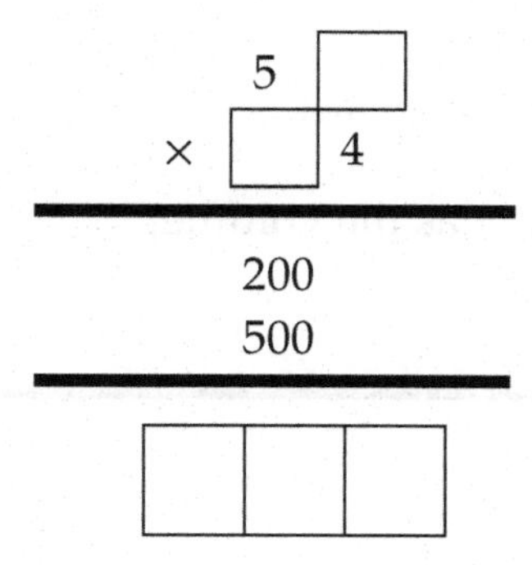

Explain your thinking to a partner.
Be able to defend your thinking in the whole group discussion.

Thursday: British Number Talk

Pick a number from each circle. Make an addition, subtraction or multiplication problem.
Write the expression under the way you solved it.

I can do it in my head.	I can do it with a model.	I can do it using a written strategy or algorithm.

A.

Write your own decimals.

B.

Write your own decimals.

Friday: Word Problem Fill-in

Fill in and solve.

The tank was longer than a yard, wider than a foot and taller than $\frac{1}{2}$ a yard.

The fish tank was ______ long, ______wide and _______tall. What was the volume?

Week 39 Teacher Notes

Monday: Missing Number

Students have to fill in the missing number.

Tuesday: Vocabulary Bingo

Students put the words on the board in any order. The teacher will play bingo. When the teacher gives the example or shows the picture students cross it out. Whoever gets 4 in a row first wins.

Wednesday: True or False?

Students read the statements and decide if they are true or false. They write that in the box. In the third column, they make their own statement.

Thursday: British Number Talk

Give students about 5 minutes to work on their own or with partners to come up with some problems, hopefully from each category. This is so they stretch. You don't want them to only choose the easy problem. Then, students should share what they did with class.

Friday: Make Your Own Problem!

Write a story about comparing 2 fractions with different denominators.

Week 39 Activities

Monday: Missing Number

a. ___ + ___ is almost 1,000

b. ___ × ___ is almost 221

c. $\frac{1}{2}$ − ___ is far from 1

d. ___ ÷ ___ is almost 50

Tuesday: Vocabulary Bingo

Put the words on the board in any order. Your teacher will play bingo. When your teacher gives the example or shows the picture you cross it out. Whoever gets 4 in a row first wins.

Words: array, height, column, row, coordinate grid, x-axis, survey, y-axis, length, width, equivalent, volume, parallel, line, coordinate, plot.

Wednesday: True or False?

Read the equations. Decide if they are true or false. Cross out the false one. Make it true. Then, make your own true or false equations. Write if they are true or false.

True or false?	Make your own and share it out.
A. $(3 \times 7) + 5 - (2 \times 5) = 16$ B. $4 \times 15 \times 5 = 4 \times 75$	

Thursday: British Number Talk

Pick a number from each circle. Decide whether to add, subtract or multiply. Write the expression under the way you solved it.

I can do it in my head.	I can do it with a model.	I can do it using a written strategy or algorithm.

A. Write your own mixed numbers.

B. Write your own mixed numbers.

Friday: Make Your Own Problem!

Write a story. Compare 2 fractions with different denominators. Tell how you can compare them.

Week 40 Teacher Notes

Monday: Number Line It!

Students have to follow the directions and plot the numbers on the number lines.

Tuesday: Vocabulary Tic Tac Toe

These are quick partner energizers. Students make up their own game by using words that they choose. Then go! They have 7 minutes to play the game. They do rock, paper, scissors to start. They take turns choosing a word and explaining it to their partner. Then, they have to do a sketch or something to show they understand the word. Everybody should play the first game, if they have time, they can play the next one. Note: it is important to call everyone back together at the end and talk about the vocabulary. Briefly go over the vocabulary.

Wednesday: Decimal of the Day

Students choose their own decimal number and fill in the chart to represent the number.

Thursday: British Number Talk

Give students about 5 minutes to work on their own or with partners to come up with some problems, hopefully from each category. This is so they stretch. You don't want them to only choose the easy problem. Then, students should share what they did with class.

Friday: Make Your Own Problem!

Students write their own word problems.

Week 40 Activities

Monday: Number Line It!

A. Plot 5 fractions on this number line.

0 2

B. Plot 5 decimals on this number line.

0 2

C. Plot 5 mixed numbers on this number line.

0 5

Tuesday: Vocabulary Tic Tac Toe

Make up your own math vocabulary game. Play with your partner.

A.

B.

Wednesday: Decimal of the Day

Choose your own decimal of the day. Fill in the boxes below based on that number.

Word form	Expanded form Expanded notation
____ is greater than ______ ____ is less than ______	____ + ____ = ______ ____ − ____ = ______ _____ × _____ almost makes ______

Thursday: British Number Talk

Pick a number from each circle. Create a problem. Solve it. Write the expression under the way you solved it.

I can do it in my head.	I can do it with a model.	I can do it using a written strategy or algorithm.

A. Make up your own numbers.

B. Make up your own numbers.

Friday: Make Your Own Problem!

Write and solve your own multi-step word problem.

Answer Key

Week 1

Monday: What Doesn't Belong

A. 20 – 9.

B. kilogram.

Tuesday: Vocabulary Match

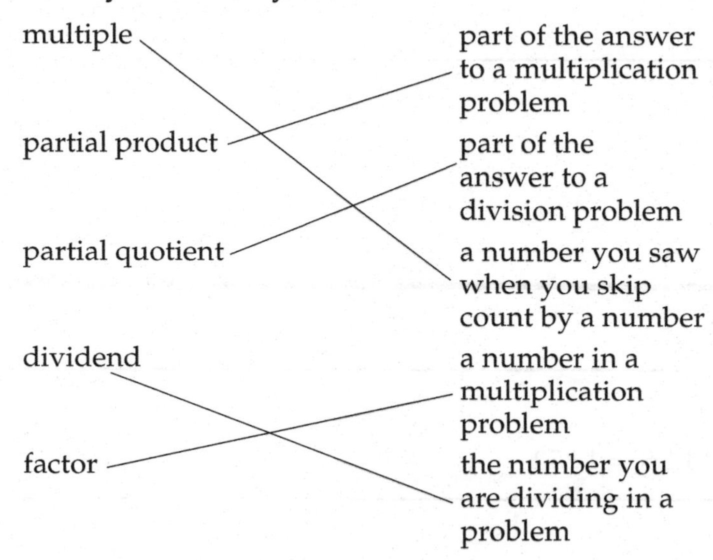

Wednesday: Guess My Number

I am greater than 57 and less than 85. I am a multiple of 7. The sum of my digits is 9. Who am I? **63**	I am greater than $\frac{3}{4}$ and less than $\frac{5}{4}$. My numerator and denominator are both composite numbers. I am equivalent to $\frac{12}{8}$ Who am I? $\mathbf{\frac{6}{4}}$

Thursday: Number Talk

Answers vary. Students should talk about adjusting 299 to 300.

Friday: What's the Question? (3 Read Protocol)

Answers vary. For example: How far did she run in total?

Week 2

Monday: Magic Square

$\frac{4}{12}$	$\frac{9}{12}$	$\frac{2}{12}$
$\frac{3}{12}$	$\frac{5}{12}$	$\frac{7}{12}$
$\frac{8}{12}$	$\frac{1}{12}$	$\frac{6}{12}$

Tuesday: Vocabulary Tic Tac Toe

Answers vary.

Wednesday: Number Line It!

$\frac{2}{6}$ $\frac{4}{12}$ $\frac{1}{2}$ $\frac{3}{5}$

Thursday: Number Talk

Answers vary. For example, 2,500 – 1,250 = 1,250.

Friday: Elapsed Time

Answers vary. For example: She could have left at 2:15 pm and came back at 3:00 p.m.

Week 3

Monday: Draw and Talk

Answers vary.

Tuesday: Frayer Model

Some answers vary.

For example: formula 2l + 2w; example: a rug can have a perimeter of 8 ft. Picture: (student sketch); Non-example: area.

Wednesday: Number of the Day

Some answers vary. $\frac{9}{8}$

Word form nine-eighths	$\frac{5}{8}+\frac{4}{8}=\frac{9}{8}$ $\frac{11}{8}-\frac{2}{8}=\frac{9}{8}$
$\frac{9}{8}$ is greater than $\frac{2}{8}$ $\frac{9}{8}$ is less than $\frac{11}{8}$ $\frac{9}{8}$ is equivalent to $\frac{18}{16}$	Decompose $\frac{1}{8}+\frac{6}{8}+\frac{2}{8}$

Thursday: Number Strings

Answers vary.

Friday: Model That!

Answers vary.

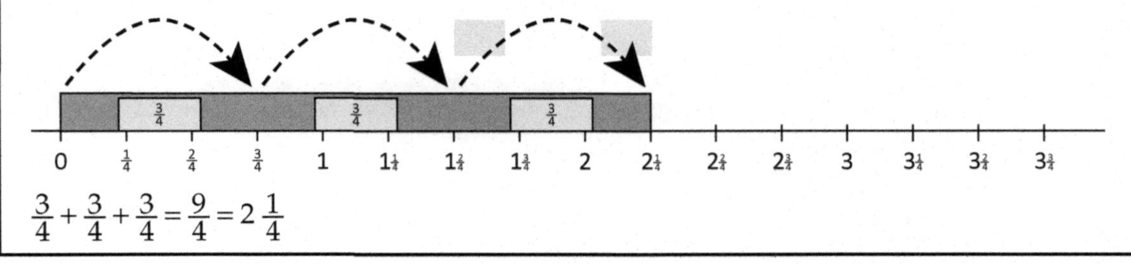

$\frac{3}{4}+\frac{3}{4}+\frac{3}{4}=\frac{9}{4}=2\frac{1}{4}$

Week 4

Monday: Convince Me!

Students should talk about associative property.

Tuesday: Frayer Model

Answers vary. Definition: remainder is the part that is left over in a division problem. Example: 4 remainder 1. Picture (drawings vary). Non-example $\frac{4}{2}$

Wednesday: Greater Than, Less Than, in Between

Answers vary.

Name a fraction greater than $\frac{1}{2}$ $\frac{3}{4}$	Name a fraction less than $\frac{2}{3}$ $\frac{1}{3}$	Name a fraction less than 1 $\frac{4}{5}$
Name a fraction greater than $\frac{1}{2}$ $\frac{5}{6}$	Name a fraction in-between $\frac{1}{2}$ and 1 $\frac{3}{5}$	Name a fraction in-between $\frac{2}{3}$ and 1 $\frac{4}{5}$

Thursday: Number Talk

Answers vary. Students should talk about adjusting the numbers so the problem becomes 1,537 – 1,400 = 137

Friday: Picture That!

Answers vary.

Week 5

Monday: 2 Arguments

Maria.

Tuesday: Talk and Draw

Answers vary.

Wednesday: Start With … Get to … By

Have students do this with a partner and then do a choral count as a class.

Thursday: British Number Talk

Answers vary.

Friday: Make Your Own Problem!

Answers vary.

Week 6

Monday: It Is/It Isn't

Answers vary. For example: it is composite. It isn't odd.

Tuesday: 1-Minute Essay

Answers vary. For example: equivalent fractions can be simplified to the same fraction. They have numerators that are multiples.

Wednesday: Find and Fix the Error

He didn't regroup. Correct answer is 822.

Thursday: Number Strings

Answers vary.

Friday: Fill in the Problem!

Answers vary.

Week 7

Monday: 3 Truths and a Fib

$7.10 < 7.01$ is false.

Tuesday: Vocabulary Brainstorm

Answers vary. Students should write and illustrate about the commutative, distributive and associative property.

Wednesday: Input/Output Table

Answers vary.

Thursday: Number Talk

Students should talk about adjusting the numbers to make an easier problem: $2 + $3.53 is an easier problem.

Friday: Get Close to

Answers vary. 10×99.

Week 8

Monday: Alike and Different

Answers vary. They both are multiples of .05.

Tuesday: Vocabulary Tic Tac Toe

Answers vary.

Wednesday: Number Bond It!

Answers vary. For example $1\frac{1}{2} + \frac{1}{3}$; $1\frac{2}{6} + \frac{3}{6}$; $\frac{1}{2} + \frac{1}{6} + 1\frac{1}{6}$

Thursday: British Number Talk

Answers vary.

Friday: What's the Story? Here's the Model

Answers vary. For example: $\frac{14}{6} = 2$ remainder 2.

Week 9

Monday: Number Line It!

Answers vary. For example: $\frac{1}{2}, \frac{3}{4}, \frac{4}{5}$

Tuesday: Vocabulary Bingo

Answers vary.

Wednesday: Fraction of the Day

$\frac{5}{4}$

Word form:	How far from 1?	Draw a model:
Five-fourths	$\frac{1}{4}$ more	
$\frac{5}{4}$ is greater than $\frac{1}{4}$ $\frac{5}{4}$ is less than $\frac{8}{2}$ $\frac{5}{4}$ is equivalent to $\frac{10}{8}$ Add a fraction that is greater than this number. Add a fraction that is less than this number. Plot an equivalent fraction on the number line. (Answers vary).		$\frac{5}{4} \times \frac{1}{2} = \frac{5}{8}$ $\frac{5}{4} - \frac{19}{16} = \frac{1}{16}$ $\frac{5}{4} + \frac{2}{8} = \frac{12}{8}$

Thursday: Number Talk Puzzle

2.97 + 5.59.

Friday: Graph Talk

Answers vary. For example: In the garden, there were several worms. 4 were $\frac{1}{2}$ an inch. 5 were an inch. 5 were $1\frac{1}{2}$ inches long. 3 were 2 inches and 7 were $2\frac{1}{2}$ inches long.

Week 10

Monday: Reasoning Matrices

$5 \times 18 = 90$
$630 = 7 \times 90$
$20 = 100 \div 5$
$20 = 1{,}000 \div 50$
$60 \times 9 = 540$.
Answers vary.

Tuesday: 1-Minute Essay

Answers vary.

Wednesday: Decimal of the Day

Some problem answers will vary.
.92

Word form Ninety-two hundredths	.10 more 1.02 .10 less .82	Round to nearest tenth .9
Expanded form .90 + .02	.40 + .40 + .12	1.00 – .92 = .08

Answers on the number line vary.

Thursday: British Number Talk

Answers vary.

Friday: What's the Story? Here's the Model

Answers vary. For example. Billy had $22.50 and he bought 5 pens. How much did each pen cost?

Week 11

Monday: Find and Fix the Error

Sara added both the numerators and the denominators. She shouldn't have added the denominators.

Tuesday: Vocabulary Bingo

Answers vary.

Wednesday: Why Is It Not?

A. Because 15.6 and 5.6 does not equal 10. B. It isn't a parallelogram because it doesn't have 2 pairs of parallel sides.

Thursday: British Number Talk

Answers vary.

Friday: What's the Question? (3 Read Protocol)

How much did Mike and Tom eat? How much did Joe eat?

Week 12

Monday: Legs and Feet

Answers vary. For example: A. 12 B. one of each and an extra cricket C. 2 cows, 2 crickets 6 chickens.

Tuesday: Vocabulary Brainstorm

Answers vary.

Wednesday: 3 Truths and a Fib

$\frac{2}{4} = \frac{5}{8}$ is the fib.

Thursday: Number Talk

Answers vary. Students should talk about making this an easier problem. \$4.22 – \$4.00 = \$.22.

Friday: What's the Problem?

Answers vary. For example: Grandma Betsy made a cake. She used $\frac{1}{6}$ cup of butter and then $\frac{3}{6}$ more. How much did she use altogether?

Week 13

Monday: Guess My Fraction

$\frac{7}{12}$

Tuesday: What Doesn't Belong?

A. addend.

B. area.

Wednesday: Fraction of the Day

Word form one and two-thirds	Picture form 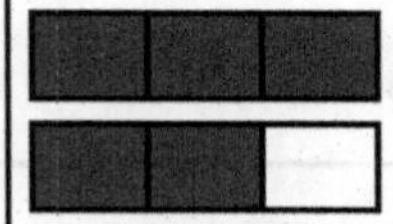
Plot it on a number line 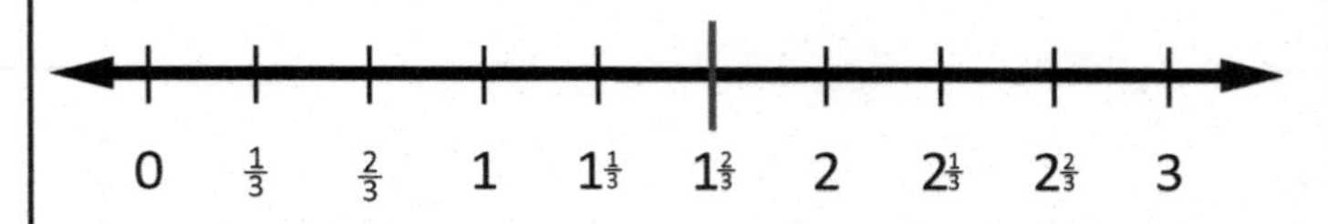	Answers vary. $1+\frac{2}{3}=1\frac{2}{3}$ $2-\frac{1}{3}=1\frac{2}{3}$ $1\frac{2}{3}\times 1=1\frac{2}{3}$

Thursday: Number Talk

Answers vary. Students talk about breaking apart numbers such as 30×71 and then $5 \times 71 = 2{,}130 + 355 = 2{,}485$.

Friday: What's the Story? Here's the Model

Answers vary. For example: The ribbon was 9.2 inches long. They wanted to make 4 bows out of the entire ribbon. How much ribbon will you use for each bow?

Week 14

Monday: Always, Sometimes, Never

A. Sometimes, it depends on the whole.

B. Sometimes.

Tuesday: Alike and Different

Answers vary. For example: decimals and fractions both represent a part of a whole.

Wednesday: Venn Diagram

A. 9, 18, 36.

B. 8, 16, 24, 48.

Overlap: 1, 2, 3, 4, 6, 12.

Thursday: British Number Talk

Answers vary.

Friday: Perimeter Problem

Answers vary. For example: it could have been 12 ft by 10 ft, 60 ft by 2 ft, 20 ft by 6 ft, 40 ft by 3 ft, 30 ft by 4 ft.

Week 15

Monday: Number Line It!

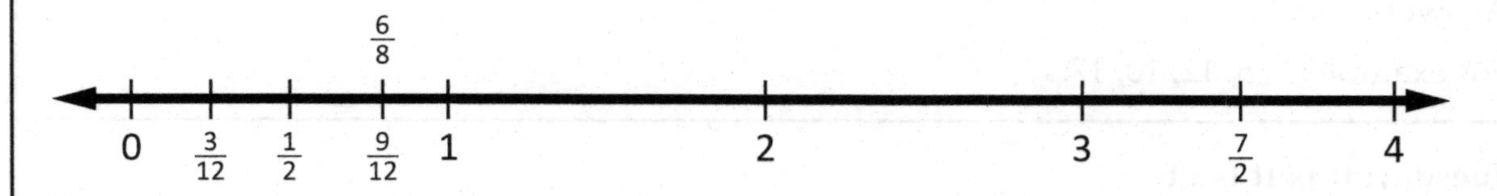

Tuesday: Vocabulary Tic Tac Toe

Answers vary.

Wednesday: Venn Diagram

Numbers less than 1: $\frac{1}{2}$, $\frac{8}{100}$, 0.09

Numbers greater than 0.8: $1\frac{1}{2}$, 1.8

Overlap: $\frac{9}{10}$, 0.9, 0.88

Thursday: British Number Talk

Answers vary.

Friday: What's the Question? (3 Read Protocol)

Answers vary. For example: How much money did they have altogether?

Week 16

Monday: Patterns/Skip Counting

Answers vary.

For example: 7, 5, 12, 10, 17.

Tuesday: It Is/It Isn't

Answers vary. It is odd. It is a factor of .25. It is not composite.

Wednesday: Convince Me!

Answers vary. Students should discuss properties.

Thursday: British Number Talk

Answers vary.

Friday: Make Your Own Problem!

answers vary. For example: Sue biked $\frac{1}{4}$ mile in the morning and $\frac{2}{8}$ of a mile in the afternoon. How far did she bike altogether? $\frac{1}{2}$ mile.

Week 17

Monday: Input/Output Table

What's the rule? **Add $\frac{2}{3}$**	What's the rule? **Halve the fraction**	Make your own fraction pattern.

In	Out
$\frac{2}{3}$	$1\frac{1}{3}$
2	$2\frac{2}{3}$
$\frac{3}{3}$	$1\frac{2}{3}$
$\frac{10}{3}$	$\frac{12}{3}$
$5\frac{1}{3}$	6

In	Out
$\frac{1}{2}$	$\frac{1}{4}$
$\frac{1}{8}$	$\frac{1}{16}$
$\frac{1}{3}$	$\frac{1}{6}$
$\frac{1}{4}$	$\frac{1}{8}$
$\frac{1}{5}$	$\frac{1}{10}$

Answers vary

Tuesday: Vocabulary Fill-in

A. sum.

B. difference.

C. quotient.

D. product.

Wednesday: What Doesn't Belong

.16 – .9

Thursday: Number Strings

Answers vary.

Friday: Equation Match

C.

Week 18

Monday: Why Is It Not?

A. $60 - 10 \times 2 = 40$, which is not 100.

B. $0.9 \times 6 = 5.4$, which is not 0.54.

Tuesday: Convince Me!

Answers vary.

Wednesday: Fraction Bingo

Answers vary.

Thursday: Number Talk

You want students to be thinking about partial quotients. For example, 600 and 90 and 10. We get $40 + 6 = 40$ with a remainder of 10.

Friday: What's the Question? (3 Read Protocol)

Answers vary. For example: How far did he run altogether?

Week 19

Monday: Alike and Different

Answers vary. For example: Students should discuss how these are all quadrilaterals. They should talk about how some are parallelograms. The one that crosses over is called a complex quadrilateral.

Tuesday: Vocabulary Bingo

Answers vary.

Wednesday: How Many More to

Students should do this with a partner and then count it out chorally with the whole class.

Thursday: British Number Talk

Answers vary.

Friday: What's the Problem?

Answers vary. For example: 2,000 divided by 25.

Week 20

Monday: True or False?

True or False	
$5 \times 7 = \frac{35}{1}$	true
$5 \times 6 \times 3 = 2 \times 30$	false
$\frac{2}{2} = \frac{10}{10}$	true
$\frac{3}{6} > \frac{1}{2}$	false
Make your own!	answers vary

Tuesday: Vocabulary Brainstorm

Answers vary.

Wednesday: Patterns/Skip Counting

Answers vary.

Thursday: Number Talk

Answers vary. Topics should include fractions that equal 1 whole, equivalent fractions and making easier problems. Students should talk about adding the same number to both parts of the subtraction problem to make an easier problem. For example $3\frac{2}{4} - 2\frac{3}{4}$. Students can make an easier problem by adding $\frac{1}{4}$ to each number and getting $3\frac{3}{4} - 3 = \frac{3}{4}$.

Friday: What's the Story? Here's the Model

Answers vary. For example: 4 friends shared $\frac{1}{2}$ a pie. They each got $\frac{1}{8}$ of the pie.

Week 21

Monday: True or False?
A. is false; B. and C. are true.

Monday: True or False?

A. is false; B. and C. are true.

Tuesday: Vocabulary Bingo

Answers vary.

Wednesday: Number Bond It!

Answers vary: 1 kilo is 500 g + 500 g. 1 kilo is 500 g + 250 g + 250 g.

Thursday: British Number Talk

Answers vary.

Friday: What's the Story? Here's the Model

Answers vary. Possible answer: Mary had $.22. At the bake sale the cookies were 5 cents each. How many can she buy? She can buy 4 cookies and she will have 2 cents left.

Week 22

Monday: Missing Numbers

$\frac{6}{8} - \frac{2}{4} = \frac{2}{8}$

Tuesday: Vocabulary Match

thousandths	.93<u>2</u>
hundredths	1.0<u>8</u>
tenths	.<u>7</u>60
whole number	5
fraction	$\frac{4}{5}$

Wednesday: Fraction of the Day

Answers vary. For example:

$2\frac{1}{2}$

<table>
<tr><td>Word form

two and one half</td><td>Picture form

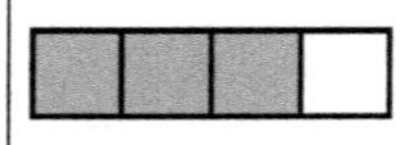</td></tr>
<tr><td>Plot it on a number line

0 $\frac{1}{2}$ 1 $1\frac{1}{2}$ 2 $2\frac{1}{2}$</td><td>$2\frac{1}{2} < 3$
$2\frac{1}{2} = \frac{5}{2}$
$2\frac{1}{2} > \frac{1}{4}$</td></tr>
</table>

Thursday: Number Talk

Answers vary. For example:

$1 + 1\frac{1}{2} = 2\frac{1}{2}$ $\quad 3 - \frac{1}{2} = 2\frac{1}{2}$ $\quad 2\frac{1}{2} \times 1 = 2\frac{1}{2}$

Friday: Graph Talk

Answers vary. Possible answer: There was a museum display of beetles. 2 were $\frac{1}{5}$ inches long. 4 were $\frac{1}{2}$ inch long. 5 were $1\frac{1}{2}$ inches long. 3 were $1\frac{3}{4}$ inches long. 6 were $2\frac{1}{8}$ inches long.

Week 23

Monday: Magic Square

3.3	2.6	3.1
2.8	3.0	3.2
2.9	3.4	2.7

Tuesday: Frayer Model

Definition A multiple of ten	Examples Money
Give a picture example A hundreds chart	Non-examples A fraction

Wednesday: Number Line It!

A. $\frac{0}{9}, \frac{7}{9}, \frac{4}{5}, \frac{7}{4}, 2\frac{1}{2}$

B. .01, .17, .34, .59, 1.02.

C. Answers vary. For example: $\frac{1}{4}, \frac{2}{3}, \frac{5}{3}, \frac{8}{4}, \frac{15}{3}$

Thursday: Number Talk

Answers vary. For example:

140 + 140 + 140 = 420
567 – 420 = 147
147 – 140 = 7
30 + 10 = 40 r7

Friday: Model That!

3 were turtles. 6 were eels and 6 were small orange fish

t	e	f

Week 24

Monday: Input/Output Table

Answers vary.

Tuesday: 1-Minute Essay

Answers vary. Possible answers: mixed numbers are a whole number and a fraction.

Wednesday: Rounding

Answers vary. .78, .81, .82.

Thursday: British Number Talk

Answers vary. $108 \div 10$ is 10.8

Friday: Model That!

The fraction of grasshoppers is 2/12 or 1/6.

4 were butterflies, 6 were ladybugs and 2 were grasshoppers.

Models vary.

Week 25

Monday: What Doesn't Belong?

A. .30 – .25.

B. (3 × 5) + (2 × 4).

Tuesday: Vocabulary Tic Tac Toe

Answers vary.

Wednesday: Decimal of the Day

Answers vary. For example:

.078

Word form seventy-eight thousandths	Answers vary. For example .068 + .010 = .078 .088 - .010 = .078 $.078 = 0 \times \frac{1}{10} + 7 \frac{1}{100} + 8 \times \frac{1}{1000}$
Answers vary.	Round it to the nearest hundredth 0.08 Round it to the nearest tenth .1

Thursday: Number Talk

Students talk about breaking apart .75. They should also think of it as money.

Friday: Make Your Own Problem!

The toy store had 112 marbles. They put them equally in 12 bags. How many marbles were in each bag? There were 9 in each bag and 4 left over. Models vary.

Week 26

Monday: Reasoning Matrices

$384 \times 1{,}000$ is more than 1,000

$0.88 \times 1{,}000$ is close to 1,000

$\frac{1}{3} \times 1{,}000$ is less than 1,000

$\frac{13}{4} \times 1{,}000$ is more than 1,000

Tuesday: Vocabulary Fill-in

- Miles and kilometers measure distance.
- We use pints, quarts, and liters to measure capacity.
- Volume measures capacity.
- Mass and weight are measured with grams and pounds.

Wednesday: Greater Than, Less Than, in Between

Answers vary. For example: .03 .45 2.79.

Name a decimal that is greater than .03 **.10**	Name a decimal greater than .45 **.78**	Name a decimal less than 2.79 **2.70**
Name a decimal that is smaller than .03 **.02**	Name a decimal in-between .03 and .45 **.15**	Name a decimal in-between .45 and 2.79 **1.50**

Thursday: Find and Fix the Error

Mike added the numerators. He should have multiplied the numerators. The answer is $\frac{1}{4}$.

Friday: What's the Story? Here's the Model

Answers vary. For example:

John has 5 marbles. Marta has twice as many as he does. Joe has 2 less than he does. How many do they have altogether?

$2 \times 5 = 10$
$5 - 2 = 3$
$5 + 10 + 3 = 18$

Week 27

Monday: 3 Truths and a Fib

A. 5 is composite is the fib. B. .90 is less than .105 is the fib.

Tuesday: Vocabulary Talk

Answers vary. For example: the numerator and denominator are parts of a written fraction.

Wednesday: Input/Output Table

Answers vary.

In	Out
12	600
10	500
2	100

Thursday: Number Talk

Answer vary. How many halves can you take out of 6? 12.

Friday: Equation Match

Problem B. Models vary.

Week 28

Monday: Why Is It Not?

It is not $\frac{5}{18}$ for several reasons. First, you can't just add numerators and denominators. You have to think about $\frac{2}{12}$ + how many more will make $\frac{3}{6}$ So we know that $\frac{2}{12}$ is equivalent to $\frac{1}{6}$ and $\frac{1}{6}$ plus $\frac{2}{6}$ makes $\frac{3}{6}$.

Tuesday: It Is/It Isn't

Answers vary.

A right angle is more than 50 degrees. It is greater than 80 degrees. It is less than 100 degrees. It is not an acute angle. It is not an obtuse angle.

Wednesday: How Many More to

A. .61

B. 0.87

C. 3.4

D. Answers vary.

Thursday: British Number Talk

Answers vary. For example $\frac{1}{2} \times \frac{1}{2} = \frac{1}{4}$

Friday: What's the Question? (3 Read Protocol)

Answers vary. For example: How many fewer lemon pies were there than strawberry ones. How many peach pies were there? How many more strawberry pies were there than peach ones?

Week 29

Monday: Guess My Number

A. .18

B. $\frac{3}{4}$

Tuesday: Vocabulary Bingo

Answers vary.

Wednesday: Find and Fix the Error

Joe did not regroup. He bunched up all his numbers on the right side of the decimal point. He has more than a whole so he must show that with the numbers.

Thursday: Number Talk

Answers vary. For example: $2\frac{1}{2} + 3$; $6 - \frac{1}{2}$; $\frac{11}{2}$

Friday: What's the Problem?

Answers vary. For example: The bakery made 65 cookies. They put them in 8 boxes. About how many were in each box. There were about 8 in each box, but 1 box had 9.

Week 30

Monday: Open Array Puzzle

720 + 64 with remainder 3. So 90 + 8 remainder 3. 98 remainder 3.

Tuesday: Vocabulary Talk

Answers vary. For example, mm is a very small measure of length. Cm is 10 mm. A meter is 100 cm and then km measure distances like driving or flying.

Wednesday: Money Mix Up

Ted had $5. He spent half on lollipops, $\frac{1}{5}$ on gum and split the rest with his little sister. Write how much money he spent on lollipops, gum, his sister and how much he kept. $2.50 on lollipops. $1 on gum and $1.50 left that he split with his sister which is $.75 each.

Thursday: British Number Talk

Answers vary. 21 divided by 7. I did it in my head. I thought of money.

Friday: What's the Problem?

Answers vary. A box was 3 inches by 5 in by 4 inches. What was the volume?

Week 31

Monday: What Doesn't Belong?

A. $7.45 = 7\frac{45}{10}$

B. $30 \times 10 \times 0$.

Tuesday: Vocabulary Brainstorm

Answers vary. Volume is how many cubic units fits into a space.

Wednesday: 3 Truths and a Fib

A. 4 is the fib.

B. 2 is the fib.

Thursday: Number Talk

Answers vary. Students should discuss the pattern.

Friday: What's the Story? Here's the Model

Answers vary. For example, Grandma ate a pie. At the end of the day there was $\frac{1}{2}$ left. 4 people still wanted a piece of pie. How much pie did everyone get?

Week 32

Monday: Open Array Puzzle

$12 \times 18 = 216$

	10 + 8	
10	100	80
+		
2	20	16

Tuesday: What Doesn't Belong?

A. prime.

B. digit.

Wednesday: 10 Times as Much

20779 – 700 hundred is 10 times more than 70.

Thursday: British Number Talk

Answers vary. For example: 81×100 is 8100.

Friday: What's the Story? Here's the Model

Grandma had $50. She gave each of her 4 grandkids the same amount of money. How much money did each kid get?

Week 33

Monday: Reasoning Matrices

A. The product will be smaller than 1.

B. The product will be greater than 1.

C. The number will stay the same.

Tuesday: Vocabulary Talk

Answers vary. For example:

Name 2 things you can measure the volume of.
Fish tanks and boxes

Name 2 things you measure the area of.
Rug, couch

Name 2 things you measure the perimeter of.
Rug, picture frame

Name 2 things that you can measure in cubic units.
Box, refrigerator

Wednesday: Decimal Comparison

Answers vary.

Thursday: Number Talk Puzzle

$31 \times 15 = 465$

Friday: Graph Talk

Answers vary. For example. The florist had several plants. 2 were $\frac{1}{2}$ inch. 1 was 1 inch. 4 were $1\frac{1}{4}$, 5 were $1\frac{1}{2}$, 3 were $1\frac{3}{4}$ and 6 were $2\frac{1}{2}$. The difference between the longest and shortest plant is 2 inches.

Week 34

Monday: Coordinate Grid

Answers vary for point: (3,1)

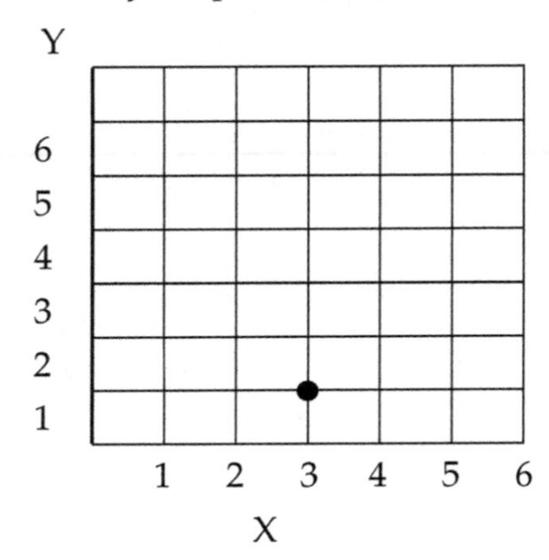

Tuesday: Vocabulary Brainstorm

Answers vary.

Wednesday: Vocabulary Fill-in

Answers vary.

- There are 1000 mm in 1 meter.
- There are 100 cl in 1 liter.
- There are 2000 g in 2 kilograms.

Thursday: Find and Fix the Error

He did not invert. He just multiplied them. He should have gotten 12.

Friday: Model That!

Answers vary. For example: perimeter was 40 ft and the area was 100 square feet.

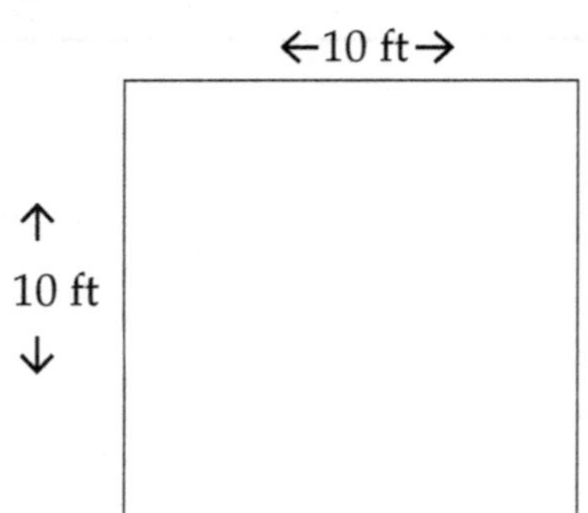

Week 35

Monday: Find and Fix the Error

Jay did not put the numbers in the correct place.

Tuesday: Vocabulary Tic Tac Toe

Answers vary.

Wednesday: Fraction of the Day

Answers vary. For example:

A. $\frac{3}{8} + \frac{4}{8} = \frac{7}{8}$

B. $3 - 1\frac{3}{8}$

C. $\frac{3}{2} \times \frac{1}{2} = \frac{3}{4}$

D. $\frac{3}{4} \times \frac{3}{4} = \frac{9}{16}$

Thursday: British Number Talk

Answers vary. For example: $1 - .4 = .6$

Friday: Perimeter Problem

The perimeter is 200 feet.

Week 36

Monday: What's the Problem?

Answers vary. Ask students do they notice a pattern.

$2.10 \div 3$; $4.90 \div 7$; $1.40 \div 2$

Tuesday: What Doesn't Belong?

A. difference.

B. gram.

Wednesday: Guess My Number

A. .77

B. .36

Thursday: British Number Talk

Answers vary.

Friday: What's the Story? Here's the Model

Story 1 is correct.

Week 37

Monday: Convince Me!

- $\frac{3}{4}$ is equivalent to $\frac{6}{8}$ and $\frac{6}{8}$ is greater than $\frac{1}{8}$
- $\frac{3}{6}$ is only $\frac{1}{2}$ but $\frac{2}{3}$ is greater than $\frac{1}{2}$ so $\frac{2}{3}$ is the larger fraction.
- $\frac{7}{6}$ is 1 and $\frac{1}{6}$ but $\frac{4}{3}$ is 1 $\frac{1}{3}$ and thirds are larger than sixths.

Tuesday: Vocabulary Bingo

Answers vary.

Wednesday: What Doesn't Belong?

A. $\frac{3}{6}$ doesn't belong.

B. $\frac{5}{4}$ is greater than 1 whole and all the other fractions are smaller than 1 whole.

Thursday: British Number Talk

Answers vary.

Friday: Make Your Own Problem!

Answers vary. For example:

A. .50 – .40

B. .75 – .25

C. 3.07 – 2.00 = 1.07

D. Sue had 50 cents and she spent 40 cents. How much does she have left? 10 cents.

Week 38

Monday: Input/Output Table

Answers vary. For example:

Input	Output
.01	.1
.04	.13

Tuesday: Vocabulary Tic Tac Toe

Answers vary.

Wednesday: Place Value Puzzle

$14 \times 50 = 700$.

Thursday: British Number Talk

Answers vary.

Friday: Word Problem Fill-in

Answers vary. For example: The fish tank was 72 inches long, 24 inches wide and 20 inches tall. The volume was 34,560 cubic inches.

Week 39

Monday: Missing Number

A. 490 + 490 is almost 1,000.
B. 10×22 is almost 221.
C. $\frac{1}{2} - \frac{3}{8}$ is far from 1.
D. $\frac{2,500}{55}$ is almost 50.

Tuesday: Vocabulary Bingo

Answers vary.

Wednesday: True or False?

A and B are true. Students write their own as well.

Thursday: British Number Talk

Answers vary.

Friday: Make Your Own Problem!

Answers vary. For example, Luke and Don had the same size candy bars. Luke ate $\frac{5}{8}$ of his candy bar. Don ate $\frac{4}{4}$ of his candy bar. Who ate more? How much more? We know that $\frac{5}{8}$ is a little more than $\frac{4}{8}$. Don ate the whole thing.

Week 40

Monday: Number Line It!

Answers vary. For example: $\frac{1}{7}, \frac{1}{4}, \frac{1}{2}, \frac{3}{4}\frac{4}{4}$; .25, .34, .55, .70, .99; $1\frac{3}{4}, 2\frac{1}{2}, 2\frac{3}{4}, 3\frac{1}{3}, 4\frac{4}{3}$

Tuesday: Vocabulary Tic Tac Toe

Answers vary.

Wednesday: Decimal of the Day

Answers vary.

Thursday: British Number Talk

Answers vary.

Friday: Make Your Own Problem!

Answers vary.

Printed in the United States
by Baker & Taylor Publisher Services

Printed in the United States
by Baker & Taylor Publisher Services